HARDWOOD FLOORS
INSTALLING, MAINTAINING AND REPAIRING

"Therefore everyone who hears these words of mine and puts them into practice is like a wise man who built his house on the rock. The rain came down, the streams rose, and the winds blew and beat against the house; yet it did not fall, because it had its foundation on the rock."

(Matthew 7:24-25, NIV)

HARDWOOD FLOORS
INSTALLING, MAINTAINING AND REPAIRING

DAN RAMSEY

TAB BOOKS Inc.
Blue Ridge Summit, PA 17214

FIRST EDITION

FIRST PRINTING

Copyright © 1985 by TAB BOOKS Inc.
Printed in the United States of America

Library of Congress Cataloging in Publication Data

Ramsey, Dan, 1945—
Hardwood floors—installing, maintaining, and
repairing.

Includes index.
1. Floors, Wooden. I. Title.
TH2529.W6R36   1985   694'.2   85-17370
ISBN 0-8306-0928-8
ISBN 0-8306-1928-3 (pbk.)

Front cover photographs courtesy of Harris-Tarkett Inc.

Contents

Acknowledgments   **vii**

Introduction   **ix**

1   Understanding Hardwood Floors   **1**

Properties of Wood—Cutting and Seasoning Lumber—Wood Flooring Materials—Wood Joints—Flooring Patterns—House Construction

2   Planning Hardwood Floors   **17**

The Floor Plan—Planning Lumber—Estimating Board Feet—Buying Hardwood Flooring—Hardwood Flooring Tools—Using Your Tape Rule—Using a Framing Square—Using a Combination Square—Handsaws—Sawing Aids—Portable Circular Saw—Safety with Power Saws—Carpenter's Level—Claw Hammer—Nailing—Nails—Wood Screws—The Foundation—Wall Framing—Flooring

3   Installing Hardwood Floors   **51**

Installation Methods—Handling and Storage—Installation Over Concrete Slabs—Installation Over Wood Joists—Laying Out the Finish Floor—Plank Flooring—Laying Over Old Flooring—Parquet and Block Flooring—Special Installation Problems—Gym Flooring—Making Installation Easier—Floating Floor Installation—Installing Plank Flooring—Installing Parquet Flooring—Adhesive—Leaving Your Mark

4   Finishing Hardwood Floors   **79**

Floor Preparation—Sanding—Preparing for the Finish—Coloring the Floor—Types of Finishes—Stenciling—Use of Fillers—Gym and Roller Rink Floors—Protecting the Finish—Base Moldings

5 Maintaining Hardwood Floors   **97**

General Care of Hardwood Floors—Daily Care—Understanding Your Finish—Maintaining Penetrating Seals—Maintaining Surface Finishes—Maintaining Polymer Finishes—Easy Floor Care—Maintaining Special Surfaces—Removing Stains—Repairing the Finish—Solving Cracks and Squeaks—Floor Repairs

6 Refinishing Hardwood Floors   **111**

When to Refinish—Refinishing—Sanding an Old Floor—New Finishes—Refinish Planing—Refinish Chiseling

Appendix A   Hardwood Floor Game Markings and Tables   **123**

Appendix B   Suppliers   **137**

Glossary   **141**

Index   **147**

Acknowledgments

Many professionals have contributed their knowledge and skills to the completeness of this book. They should be thanked: National Oak Flooring Manufacturers Association; Maple Flooring Manufacturers Association; American Parquet Association, Inc.; Wood and Synthetic Flooring Institute; Harris-Tarkett, Inc.; Lavidge & Associates, Inc.; Pennwood Products Co.; The Jennison-Wright Corp.; Dixon Lumber Company, Inc.; Oregon Lumber Company; U.S. Department of Agriculture, Forest Service; Department of the Army; and Department of the Navy. Thank you!

Introduction

Hardwood floors, popular a few decades ago, are returning to new and remodeled homes for both fashion and function. Meanwhile, older homes are being restored, and seasoned hardwood floors reconditioned.

Hardwood Floors—Installing, Maintaining and Repairing introduces the benefits of wood flooring to the new consumer. It thoroughly covers the planning, selection, installation, finishing, maintenance, repair, and refinishing of all types of hardwood flooring. Tongue-and-groove strip flooring is the most popular. Also covered are planking, block flooring, and parquet floors. Step-by-step instructions and illustrations show you how to tackle every aspect of hardwood floor installation and repair like a professional.

Chapter 1

Understanding Hardwood Floors

W OOD FLOORING HAS BEEN A TRADITIONAL American favorite since pioneer homes were first erected from that most readily available resource. Today, however, hardwood flooring is expensive. The selection and installation of quality hardwood flooring often requires greater expense than do other common flooring materials. Even so, the lower maintenance costs and the aesthetic qualities of hardwood flooring, combined with reduced costs to the do-it-yourselfer, have spurred a renewed interest in practical hardwood floors. An important reason is the natural beauty of hardwood floors (Figs. 1-1 through 1-3), which blends with all types and tastes of interior decorating, from traditional to contemporary.

To guide you in understanding hardwood flooring (Fig. 1-4), let's consider the properties of wood, types of wood joints, common types of hardwood flooring, and how they fit into modern home construction.

PROPERTIES OF WOOD

Any piece of wood is made up of a number of small cells as shown in Fig. 1-5. The size and arrangement of the cells determine the grain of the wood and many of its properties. Examine a freshly cut tree stump and you'll see that the millions of small cells are arranged in circular rings around the *pith*, or center, of the tree (Fig. 1-6). These rings are caused by a difference in the rate of growth of the tree during the various seasons of the year. In spring a tree grows rapidly and builds up a thick layer of comparatively soft, large cells that appear in the cross section of the trunk as the light-colored annual rings.

As the weather becomes warmer during the early summer, the rate of growth slows, and the summer growth forms cells that are more closely packed. These pairs of concentric springwood and summerwood rows form the annual rings, which can be counted to find out the age of the tree. Because of climatic conditions, some trees, such as oak and walnut, have more distinctive rings than others, such as maple and birch. White pine is so uniform that you can hardly distinguish the rings, while many other softwoods have a very pronounced contrast

Fig. 1-1. Hardwood floors offer natural beauty (courtesy National Oak Floor Manufacturers Association).

Fig. 1-3. Hardwood floors blend well with all types of decor (courtesy National Oak Floor Manufacturers Association).

Fig. 1-2. Hardwood floors especially complement traditional furniture (courtesy National Oak Floor Manufacturers Association).

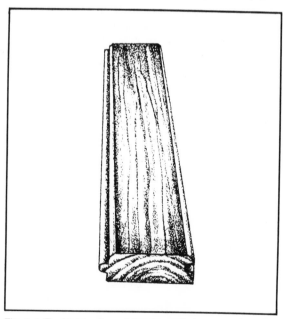

Fig. 1-4. Typical hardwood flooring strip (courtesy National Oak Floor Manufacturers Association).

2

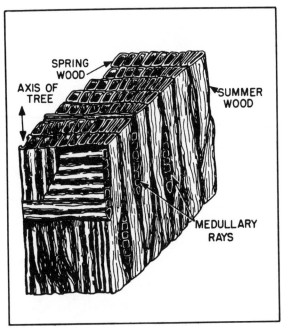

Fig. 1-5. The structure of wood.

it becomes darker. Depending upon the type of tree, it requires from 9 to 36 years to transform sapwood into heartwood.

The *cambium layer,* the boundary between the sapwood and the bark, is the thin layer where new sapwood cells form. *Medullary rays* are radial lines of wood cells consisting of threads of pith which serve as lines of communication between the central cylinder of the tree and the cambium layer. They are especially prominent in oak.

When a tree is sawed lengthwise, the annual ring forms a pattern, called the *grain* of the wood. Many terms are used to describe the various grain conditions. If the cells that form the grain are closely packed and small, the wood is said to be *fine-grained,* or *close-grained.* Maple and birch are excellent examples of this type of wood. If the cells are large, open, and porous, the wood is *coarse-grained,* or *open-grained,* as in oak, walnut, and mahogany. Furniture and flooring made of open-grained woods require the use of wood filler to close the pores and provide a smooth outside finish.

When the wood cells and fibers are comparatively straight and parallel to the trunk of the tree, the wood is said to be *straight-grained.* If the grain is crooked, slanted, or twisted, it is said to be *cross-grained.* It is the arrangement, direction, size, and color of the wood cells that give the grain of each wood its characteristic appearance.

between summerwood and springwood, making it easy to distinguish the rings.

The *sapwood* of a tree is the outer section of the tree between the *heartwood* (darker center wood) and the *bark.* The sapwood is lighter in color than the heartwood, but as it gradually changes to heartwood on the inside and as new layers are formed,

Fig. 1-6. Cross section of a tree.

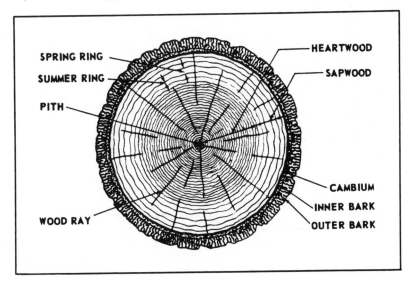

3

CUTTING AND SEASONING LUMBER

In large lumber mills, such as those found in the Pacific Northwest, logs are usually processed into lumber with huge band or circular saws. There are two methods of sawing up logs: slash cutting and rift cutting (Fig. 1-7). *Slash cutting* is accomplished by a series of cuts parallel to the side of the log. If hardwoods are being cut, the process is known as *plain sawing*. If softwoods are being cut, the process is referred to as *flat-grain sawing*.

Lumber that is specially cut to provide edge grain on both faces is said to be *rift-cut*. If hardwood is being cut, the lumber is said to be *quartersawed* (Fig. 1-8). If softwood is being sawed, it is called *edge-grain lumber*. Incidentally, if an entire log is slash-cut, several boards from near the center of the log will actually be rift-cut.

Slash-cut lumber is usually cheaper than rift-cut lumber because it takes less time to slash-cut

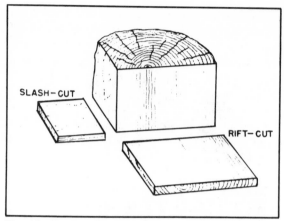

Fig. 1-7. Slash and rift cutting.

a log, and there is less waste. Circular or oval knots appearing in slash-cut boards affect the strength and surface appearance much less than do spike knots, which may appear in rift-cut boards. If a log is sawed

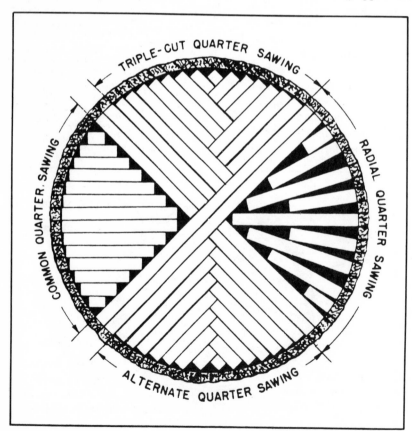

Fig. 1-8. Four methods of quartersawing.

to produce all slash-cut lumber, however, more boards will contain knots than if the log were sawed to produce the maximum amount of rift-cut material. Another advantage of slash cutting is that shakes and pitch pockets, when present, will extend through fewer boards.

For many applications, especially flooring, rift-cut lumber is preferred because it offers more resistance to wear than does slash-cut lumber. Rift-cut lumber also shrinks and swells less in width. Another advantage of rift-cut lumber is that it twists and cups less and splits less when used than does slash-cut lumber. Rift-cut lumber usually holds paint and other finishes better, also.

After being sawed, lumber must be thoroughly dried before it is suitable for most uses. The old method—and one still preferred for some uses—was merely to air-dry the lumber in a shed or stack it in the open. This method requires considerable time—up to 7 years for some of the hardwoods.

A faster method is known as *kiln drying.*. The wood is placed in a tight enclosure, called the *kiln,* and dried with heat supplied by artificial means. The length of time required for drying varies from two or three days to several weeks, depending on the kind of wood, its dimensions, and the method of drying.

Lumber is considered dry enough for most uses when the moisture content has been reduced to about 12 or 15 percent. If you use lumber very much, you will soon learn to judge the dryness of wood by color, weight, smell, feel, and by a visual examination of shavings and chips. Your lumber supplier can also give you a close estimate of dryness.

Briefly, seasoning of lumber is accomplished by removing the moisture from the millions of small and large cells of which wood is composed. Moisture (water or sap) occurs in two separate forms: free water and embedded water. *Free water* is the amount of moisture the individual cells contain. *Embedded water* is the moisture absorbed by the cell walls.

During drying or seasoning, the free water in the individual cells evaporates until a minimum amount of moisture is left. The point at which this minimum moisture remains is called the *fiber satura-tion point.* The moisture content of this point varies from 25 to 30 percent. Below the fiber saturation point, the embedded water is extracted from the porous cell walls. This process causes a reduction of the thickness of the walls. Wood shrinks across the grain when the moisture content is lowered below the fiber saturation point.

Shrinking and swelling of the wood cells caused by varying amounts of moisture change the size of the cells. Therefore, lowering or raising the moisture content causes lumber to shrink or swell. The loss of moisture during seasoning causes wood to be harder, stronger, stiffer, and lighter in weight— all qualities important to the life of hardwood flooring.

WOOD FLOORING MATERIALS

There is a wide selection of wood materials that may be used for flooring. Hardwoods and softwoods are available as strip flooring (Fig. 1-9) in a variety of widths and thicknesses, and as random-width planks and block flooring.

Softwood finish flooring costs less than most hardwood species and is often used to good advantage in bedroom and closet areas where traffic is light. It might also be selected to fit the interior decor. It is less dense than the hardwoods and less wear-resistant, and shows surface abrasions more readily. Softwoods most used for flooring are southern pine, Douglas fir, redwood, and western hemlock. While this book primarily covers hardwood flooring, many of the principles and instructions also apply to softwood flooring.

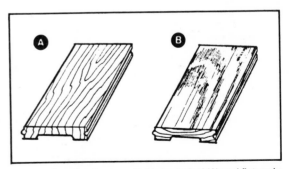

Fig. 1-9. Strip flooring is available in vertical (A) and flat-grain (B) hardwood and softwood stock.

Table 1-1 lists the grades and description of softwood strip flooring. Softwood flooring has tongued-and-grooved edges and may be hollow-backed or grooved. Some types are also end-matched. Vertical-grain flooring generally has better wearing qualities than flat-grain flooring under hard usage.

Hardwoods most commonly used for flooring are red and white oak, beech, birch, maple, and pecan. Table 1-1 also lists grades, types, and sizes of hardwood strip flooring. Manufacturers supply both prefinished and unfinished flooring.

Perhaps the most widely used pattern is a 25/32-×-2 1/4-inch *strip flooring.* These strips are laid lengthwise in a room and normally at right angles to the floor joists. Some type of a subfloor of diagonal boards or plywood is normally used under the finish floor. Strip flooring of this type is tongued-and-grooved and end-matched (Fig. 1-10). Strips are of random length and may vary from 2 to 16 feet or more. End-matched strip flooring in 25/32-inch thickness is generally hollow-backed. The face is slightly wider than the bottom so that tight joints result when flooring is laid. The tongue fits snugly into the groove to prevent movement and floor squeaks. All of these details are designed to provide beautiful finished floors that require a minimum of maintenance.

Another matched pattern may be obtained in 3/8-×-2-inch (Fig. 1-11). It is commonly used for remodeling work or when the subfloor is edge-blocked or thick enough to provide very little deflection under loads.

Square-edged strip flooring (Fig. 1-12) might also be used occasionally. It is usually 3/8 × 2 inches in size and is laid up over a substantial subfloor. Face nailing is required for this type.

Wood block flooring (Fig. 1-13) is made in a number of patterns. Blocks may vary in size from 4 × 4 inches to 9 × 9 inches and larger. Thickness varies by type from 25/32 inch for laminated blocking or plywood block tile to 1/8-inch stabilized veneer. Solid wood tile is often made up of narrow strips of wood splined or keyed together in a number of ways. Edges of the thicker tile are tongued-and-grooved, but thinner sections of wood are usually square-edged (Fig. 1-14). Plywood blocks may be 3/8 inch and thicker and are usually tongued-and-grooved (Fig. 1-15). Many block floors are factory-finished and require only waxing after installation (Fig. 1-16).

Table 1-1. Wood Flooring Grades and Graining.

Species	Grain orientation	Size		First grade	Second grade	Third grade
		Thickness	Width			
		In.	*In.*			
			SOFTWOODS			
Douglas-fir and hemlock	Edge grain	3/4	1 1/8–5 1/8	B and Better	C	D
	Flat grain	3/4	1 1/8–5 1/8			
Southern pine	Edge grain and Flat grain	5/16–1 1/4	1 1/8–5 1/8	B and Better	C, C and Better	D (and No. 2)
			HARDWOODS			
Oak	Edge grain	3/4	1 1/2–3 1/4	Clear	Select	---------------
	Flat grain	11/32	1 1/2, 2	Clear	Select	No. 1 Common
		15/32	1 1/2, 2			
Beech, birch, maple, and peacan [1]		25/32	1 1/2–3 1/4	First grade	Second grade	Third grade
		3/8	1 1/2, 2 1/4			
		1/2	1 1/2, 2 1/4			

[1] Special grades are available in which uniformity of color is a requirement.

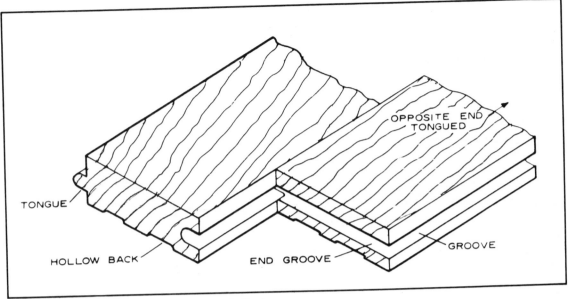

Fig. 1-10. Tongue-and-groove, end-matched strip flooring.

Figure 1-17 illustrates a typical installation of a tongue-and-groove parquet block floor unit. Figure 1-18 illustrates cross matching parquet tiles.

WOOD JOINTS

Another important element of hardwood flooring is the way wood strips or blocks are joined together. I've already mentioned the tongue-and-groove joint. Other types of joints used in woodworking include the butt, lap, miter, rabbet, dado, gain, mortise-and-tenon, slip tenon, box corner, and dovetail. Few are ever used, however, for the manufacturing and installation of hardwood floors.

Lap joints (Fig. 1-19) can be used for hardwood flooring. but are usually not. The more common types of lap joints are the plain lap, cross half-lap, end butt half-lap, and corner half-lap.

Figure 1-20 illustrates the simplest of joints—the plain butt joint. It is made by butting or placing two pieces of wood together. As mentioned earlier, but joints must be *face-nailed*, or nailed from the face and through the wood, in order to be fastened securely.

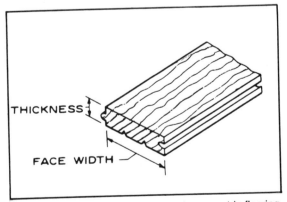

Fig. 1-11. Matched-pattern tongue-and-groove strip flooring.

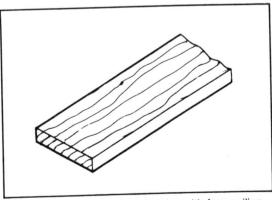

Fig. 1-12. Square-edged strip flooring with face nailing.

7

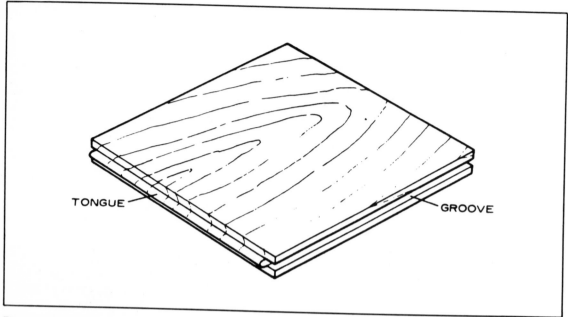

Fig. 1-13. Wood block flooring.

Figure 1-21 shows one way of overcoming this problem. Holes are drilled and connecting dowels installed to fasten members together. This is a doweled joint.

Taking the doweled joint one step further is the spline joint (Fig. 1-22). A notch is cut in both members and a smaller member, called a *spline,* is placed between them. The milling and insertion of a spline joint takes time and slows down the installation of the flooring.

The most popular joint for flooring is called the tongue and groove (Fig.1-23). The spline is actually milled from part of one of the wood members. The spline is then called a *tongue* and is designed to fit into a groove—hence the names. This type of joint offers rigid connection between flooring pieces with-

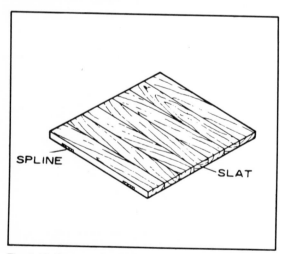

Fig. 1-14. Square-edged and splined wood block flooring.

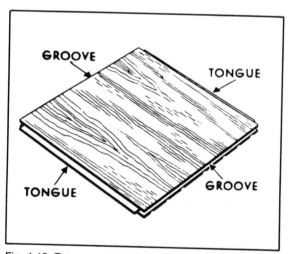

Fig. 1-15. Tongue-and-groove laminated block flooring.

8

Fig. 1-16. Factory-finished wood block flooring (courtesy Pennwood Products Co.).

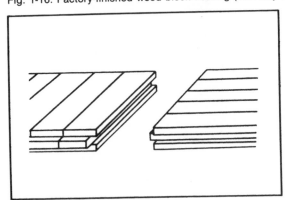

Fig. 1-17. Tongue-and-groove parquet block floor unit (courtesy Pennwood Products Co.).

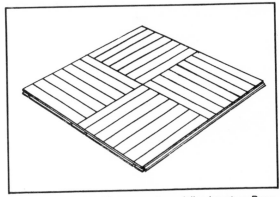

Fig. 1-18. Cross-matching parquet wood tiles (courtesy Penn-wood Products Co.).

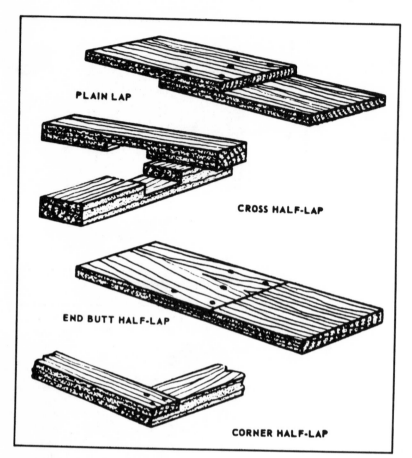

PLAIN LAP

CROSS HALF-LAP

END BUTT HALF-LAP

CORNER HALF-LAP

Fig. 1-19. Four types of lap joints.

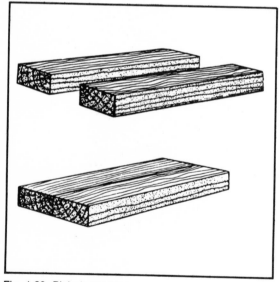

Fig. 1-20. Plain butt joints.

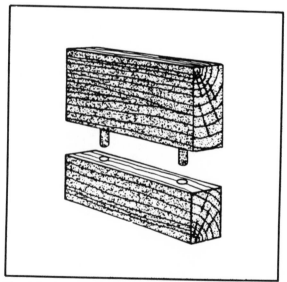

Fig. 1-21. Doweled joint.

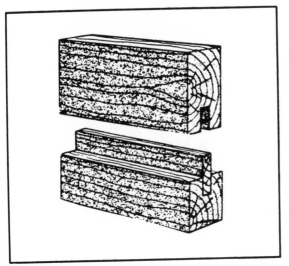

Fig. 1-22. Spline joint.

Fig. 1-23. Tongue-and-groove joint.

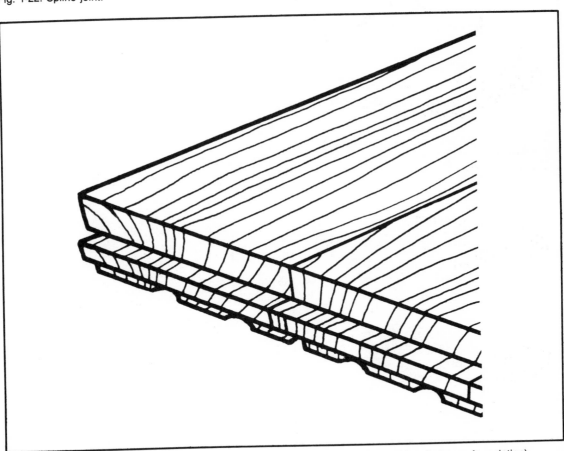

Fig. 1-24. Typical manufactured hardwood flooring joint (courtesy Maple Flooring Manufacturers Association).

Fig. 1-25. Identical length strip flooring (courtesy Pennwood Products Co.).

out excessive time for assembly.

Figure 1-24 illustrates one popular type of flooring joint that is actually a combination of joints. As you can see, the sides are butt-jointed. The edge is a tongue-and-groove joint. If installed tightly, this combined joint can be quicker and just as tight as a full tongue-and-groove joint on all four sides.

FLOORING PATTERNS

There are numerous ways you can install hardwood flooring in various patterns that will make your floor unique as well as functional. Figure 1-25 illustrates how strip flooring can be installed using boards of the same length rather than of random lengths.

Fig. 1-26. Herringbone design hardwood floor (courtesy Pennwood Products Co.).

Figure 1-26 shows a diagonal cross or herringbone design that is easy to install. It can be of identical strip flooring or of patterned block flooring.

Figures 1-27 through 1-35 illustrate a variety of block flooring patterns that can be found in many flooring departments or stores. They offer an infinite number of combinations that can beautify your home while saving you the labor of assembling and installing such intricate patterns.

Fig. 1-29. Monticello design block flooring (courtesy Pennwood Products Co.).

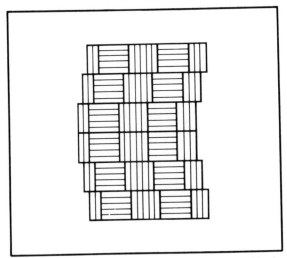

Fig. 1-27. Staggered block flooring pattern (courtesy Harris-Tarkett, Inc.).

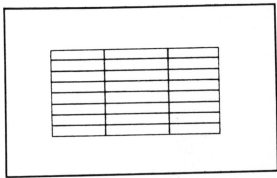

Fig. 1-30. One-directional block flooring design (courtesy Pennwood Products Co.).

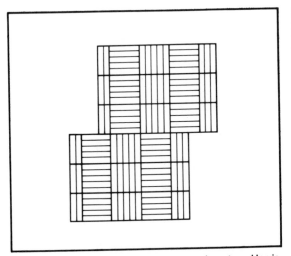

Fig. 1-28. Overlap block flooring pattern (courtesy Harris-Tarkett, Inc.).

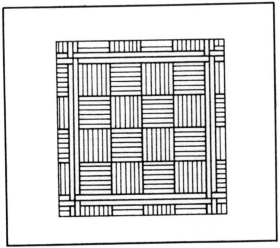

Fig. 1-31. Log cabin design block flooring (courtesy Pennwood Products Co.).

Fig. 1-32. Herringbone design block flooring (courtesy Penn-wood Products Co.).

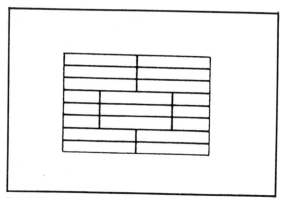

Fig. 1-33. Ashlar design block flooring (courtesy Pennwood Products Co.).

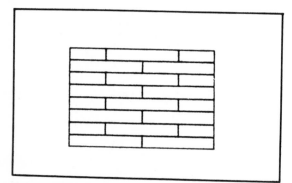

Fig. 1-34. Brick design block flooring (courtesy Pennwood Products Co.).

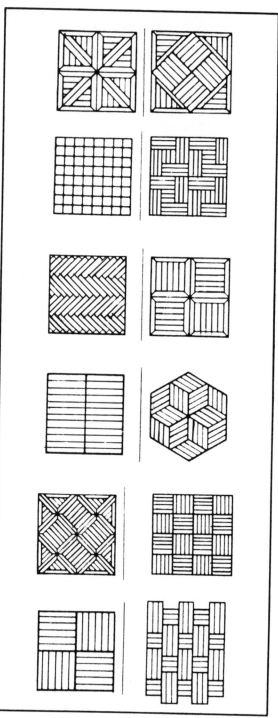

Fig. 1-35. Assorted parquet hardwood flooring designs (courtesy Pennwood Products Co.).

HOUSE CONSTRUCTION

Figure 1-36 is an exploded view of a single story, wood frame house showing the major parts. The floor system, interior and exterior walls, and the roof are the major components of such a house. Houses with flat or low-sloped roofs are usually variations of these systems.

The illustration shows a floor system con- structed over a crawl space. Supporting beams are fastened to treated posts embedded in soil or to masonry piers or a foundation. Floor joists are fastened to these beams, and the subfloor nailed to the joists, resulting in a level, sturdy platform upon which the rest of the house is constructed. This type of construction is often called *platform construction*.

Exterior walls, often assembled flat on the

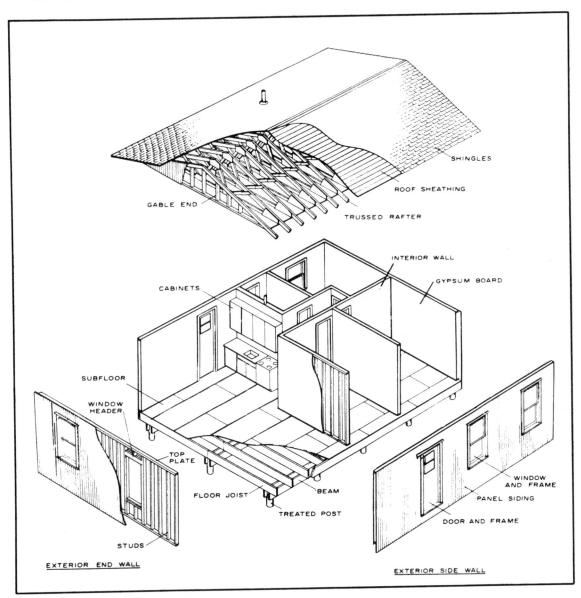

Fig. 1-36. Exploded view of a wood-frame home.

subfloor and raised in "tile-up" fashion, are fastened to the perimeter of the floor platform. Exterior coverings and window and door units are included after walls are plumbed and braced.

Interior walls are usually the next components to be erected unless trussed rafters (roof trusses) are used. Trussed rafters are designed to span from one exterior sidewall to the other and do not require support from interior partitions allowing partitions to be placed as required for room dividers. When ceiling joists and rafters are used, a bearing partition near the center of the width is necessary.

Several systems can be used to provide a roof over the house. One consists of normal ceiling joists and rafters that require some type of load-bearing wall between the sidewalls. Another is the trussed rafter system, commonly called trusses. This design requires no load-bearing walls between the sidewalls. A third design consists of thick wood roof decking. A fourth is open beams and decking that span between the exterior walls and a center wall or ridge beam. The truss and the conventional joist-and-rafter construction require some type of finish and as a surface to which the roofing material is applied.

Figure 1-37 shows the typical components of a

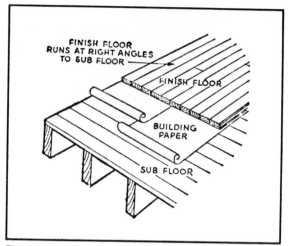

Fig. 1-37. Laying out a hardwood floor over a subfloor.

hardwood floor. The joists support the subfloor. Building paper is laid down to prevent moisture from entering the flooring from below and possibly damaging the finish. Finally, finish flooring is installed, often at right angles to the subfloor.

Many of these construction terms will be new to you. The glossary at the back of the book defines these terms. Refer to it often; it will help you understand working with hardwood flooring.

Chapter 2

Planning Hardwood Floors

A GOOD JOB BEGINS WITH GOOD PLANNING. This adage is especially true for installing and maintaining hardwood flooring. Many installation problems can be solved on paper before work begins, saving time and materials. In this chapter you'll learn how to plan and prepare for your hardwood flooring project by selecting the correct materials, properly choosing and using tools and fasteners, and learning how hardwood flooring can be used in your home.

THE FLOOR PLAN

Figure 2-1 illustrates a *floor plan*, or building plan. In home construction, the floor plan guides the contractor and subcontractors in the erection of walls, floors, and roof. Information on a floor plan includes the lengths, thicknesses, and character of the building walls at that particular floor; the widths and locations of door and window openings; the lengths and character of partitions; the number and arrangement of rooms, and the types and locations of utility installations. In many cases, the floor plan

will also establish the type and layout of flooring.

A floor plan of your home can be valuable to you when you estimate, plan, and install hardwood flooring. You may have a floor plan for your home from the contractor or from an appraiser, or you may have to generate one yourself with a tape measure and some graph paper. In either case, you'll find the time spent on researching your home's floor plan worthwhile as your hardwood flooring project continues.

PLANNING LUMBER

Hardwood flooring is lumber. You may also use other types of structural lumber as you install and repair hardwood flooring: subflooring, floor joists, sole plates, wall studs, and headers. Let's consider the planning and selection of lumber.

Lumber is usually sawed into standard size, length, width, and thickness, which permits uniformity in planning structures and in ordering materials. Table 2-1 lists the common widths and thicknesses of wood in rough and in dressed dimen-

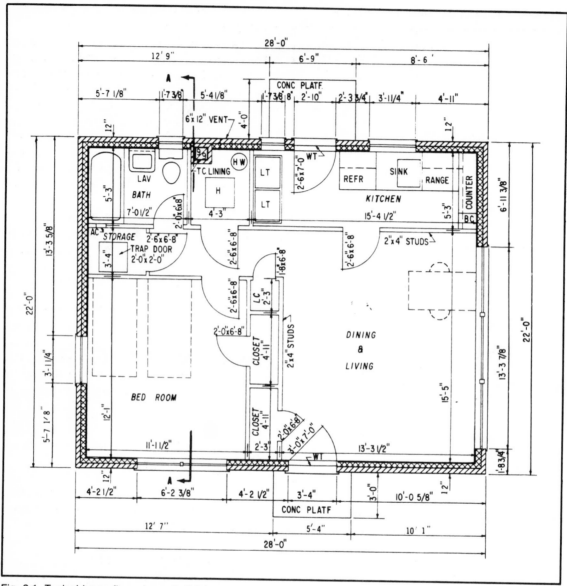

Fig. 2-1. Typical home floor plan or building plan.

sions in the United States. Standards have been established for dimension differences between the quoted size of lumber and its standard sizes when dressed. Quoted sizes refer to dimensions prior to surfacing. These differences must be taken into consideration.

A good example of the dimension difference is the common 2-×-4. As you can see in the table, the familiar quoted size of 2 × 4 is the rough or nominal dimension, while the actual dressed size is 1 5/8 × 3 5/8 inches.

Lumber, as it comes from the sawmill, is divided into three main classes: yard lumber, structural material, and factory and shop lumber. *Yard lumber* is that used for ordinary construction and general building purposes. It is subdivided into

Table 2-1. Nominal and Standard Lumber Sizes.

Nominal size (in.)	American standard (in.)
1 x 3	25/32 x 2 5/8
1 x 4	25/32 x 3 5/8
1 x 6	25/32 x 5 5/8
1 x 8	25/32 x 7 1/2
1 x 10	25/32 x 9 1/2
1 x 12	25/32 x 11 1/2
2 x 4	1 5/8 x 3 5/8
2 x 6	1 5/8 x 5 5/8
2 x 8	1 5/8 x 7 1/2
2 x 10	1 5/8 x 9 1/2
2 x 12	1 5/8 x 11 1/2
3 x 8	2 5/8 x 7 1/2
3 x 10	2 5/8 x 9 1/2
3 x 12	2 5/8 x 11 1/2
4 x 12	3 5/8 x 11 1/2
4 x 16	3 5/8 x 15 1/2
6 x 12	5 1/2 x 11 1/2
6 x 16	5 1/2 x 15 1/2
6 x 18	5 1/2 x 17 1/2
8 x 16	7 1/2 x 15 1/2
8 x 20	7 1/2 x 19 1/2
8 x 24	7 1/2 x 23 1/2

classifications of select lumber and common lumber.

Select lumber is of good appearance and finish. It is identified by the following grade names: grade A, grade B, grade C, and grade D.

Common lumber is suitable for general construction and utility purposes. It has the following grade names: No. 1 common, No. 2 common, No. 3 common, No. 4 common, and No. 5 common.

Softwood flooring is graded as select: A, B, C, and D. First-grade softwood flooring is B and better, with second-grade as C or C and better.

Hardwood flooring is graded as Clear, Select, then No. 1 common, etc. Some types of hardwood will be graded as first-grade, second-grade, etc.

ESTIMATING BOARD FEET

Sizes of lumber and woods are standardized for convenience in ordering and handling. Building material and finish material sizes run 8, 10, 12, 14, 16, 18, and 20 feet in length; 2, 4, 6, 8, 10, and 12 inches in width; and 1, 2, and 4 inches in thickness. The actual width and thickness of dressed lumber are considerably less than the standard, or quoted, width and thickness. Hardwoods, which have no standard lengths or widths, run 1/4, 1/2, 1, 1 1/4, 1 1/2, 2, 2 1/2, 3, and 4 inches in thickness.

Plywoods are usually 4 × 8 feet and vary in thickness from 1/8 to 1 inch. Stock panels are usually available in widths of 48 inches and lengths varying in multiples of 16 inches up to 8 feet. Panel lengths run in 16-inch multiples because the accepted spacing for studs and joists is 16 inches.

The amount of lumber required is measured in board feet. A *board foot* is a unit measure representing an area of 1 square foot and a thickness of 1 inch actual or nominal size. The number of board feet in a piece of lumber can be computed by the arithmetic method or by using a table.

To determine the number of board feet in one or more pieces of lumber, the following formula is used:

$$\frac{\text{Pieces} \times \text{Thickness} \times \text{Width} \times \text{Length}}{12}$$

where the thickness and width are expressed in inches and the length in feet. As an example, here's how to find the number of board feet in a piece of lumber 2 inches thick, 10 inches wide, and 6 feet long:

$$\frac{1 \times 2 \times 10 \times 6}{12} = 10 \text{ board feet}$$

Rapid estimation of board feet can also be made using Tables 2-2 or 2-3.

BUYING HARDWOOD FLOORING

As with most consumer products, how you purchase can be as important to the job as what you

Table 2-2. Rapid Calculation of Board Measure.

Width	Thickness	Board feet
3″	1″ or less	1/4 of the length
4″	1″ or less	1/3 of the length
6″	1″ or less	1/2 of the length
9″	1″ or less	3/4 of the length
12″	1″ or less	Same as the length
15″	1″ or less	1 1/4 of the length

purchase. If you buy inferior hardwood flooring and don't select a retailer who will stand behind what is sold, you can't expect a superior-quality floor.

An important and often overlooked fact is that when you buy any product, you are also buying whatever support service the retailer offers. At the big chain store that has some hardwood flooring on sale, you will probably know more about the product you buy than will the clerk who sells it to you. Unless you are an expert in the product you are selecting, avoid such retailers. You will be farther ahead to buy from a retailer who can support merchandise with product knowledge and assistance.

How can you purchase hardwood flooring from specialized retailers and still get a good price? First, select a retailer from whom you would feel comfortable buying if you had more money. Then let them know that you don't have to buy it immediately and that you would prefer to wait until you could earn a savings of 25 percent or more. Most retailers understand and will either work out some type of discount nearing or surpassing this figure, or will suggest when that price discount will be available.

To make the best purchase, you must be sure there are no "hidden costs" and know exactly what is included with your purchase. Once you know which type of hardwood flooring to select, how it will be installed, and approximately how much you will need, make a list of components to buy—flooring, fasteners, adhesives, tools, finishes, special equipment rentals. Then you can make sure that the "total price" really is the total.

You also need to know what you're buying. You should know the characteristics of wood and understand the different types of joints and flooring patterns—information offered in Chapter 1. Later in this chapter you will find out about tools and fasteners. Read about them before you go on your floor shopping trip.

One method used by smart consumers to ensure that what they bought is what they thought they bought is to defer partial payment. Make arrangements to pay part of the bill when sold, part on delivery after an inspection and inventory, and the final amount 30 days after delivery. The floor-

ing retailer who often works with local contractors is accustomed to such terms, and these terms offer you the opportunity to verify the quantity and quality of hardwood flooring and related materials with recourse.

Remember that a bargain is only a bargain when the job is complete. What may have initially cost less to purchase may eventually cost more. One of the reasons you are doing it yourself is to save money. By purchasing your hardwood flooring as a smart consumer you will both save money and have a better quality floor when you're done.

HARDWOOD FLOORING TOOLS

There are a variety of hand and power tools available to the do-it-yourselfer who wants to install hardwood flooring inexpensively and efficiently. Figure 2-2 illustrates the basic tools suggested by one hardwood flooring manufacturer. They include a handsaw or power saw, hammer, crowbar, square, chalk line, wedge, adhesive, and a special installation tool.

One of the most important tools for hardwood floor installation is the measuring tool. The ability to lay out and measure accurately depends upon the correct use of measuring tools and the ease with which the graduations on these tools are read. While each measuring tool is used for a specific purpose, they are all graduated according to the same system of linear measure.

Figure 2-3 shows types of rules and tapes commonly used by builders and do-it-yourselfers. Of all measuring tools, one of the most practical is the steel rule. This rule is usually 6 or 12 inches in length. Steel rules may be flexible or nonflexible, but the thinner the rule, the easier it is to measure accurately because the division marks are closer to the work.

Generally, a rule has four sets of graduations, one on each edge of each side. The longest lines represent the inch marks. On one edge, each inch is divided into eight equal spaces; each space represents 1/8 inch. The other edge of this side is divided into sixteenths. The 1/4-inch and 1/2-inch marks are commonly made longer than the smaller division marks to facilitate counting. The gradua-

Table 2-3. Board Feet.

Nominal size (in.)	Actual length in feet								
	8	10	12	14	16	18	20	22	24
1 x 2		1 2/3	2	2 1/3	2 2/3	3	3 1/2	3 2/3	4
1 x 3		2 1/2	3	3 1/2	4	4 1/2	5	5 1/2	6
1 x 4	2 3/4	3 1/3	4	4 2/3	5 1/3	6	6 2/3	7 1/3	8
1 x 5		4 1/6	5	5 5/6	6 2/3	7 1/2	8 1/3	9 1/6	10
1 x 6	4	5	6	7	8	9	10	11	12
1 x 7		5 5/8	7	8 1/6	9 1/3	10 1/2	11 2/3	12 5/6	14
1 x 8	5 1/3	6 2/3	8	9 1/3	10 2/3	12	13 1/3	14 2/3	16
1 x 10	6 2/3	8 1/3	10	11 2/3	13 1/3	15	16 2/3	18 1/3	20
1 x 12	8	10	12	14	16	18	20	22	24
1 1/4 x 4		4 1/6	5	5 5/6	6 2/3	7 1/2	8 1/3	9 1/6	10
1 1/4 x 6		6 1/4	7 1/2	8 3/4	10	11 1/4	12 1/2	13 3/4	15
1 1/4 x 8		8 1/3	10	11 2/3	13 1/3	15	16 2/3	18 1/3	20
1 1/4 x 10		10 5/12	12 1/2	14 7/12	16 2/3	18 3/4	20 5/6	22 11/12	25
1 1/4 x 12		12 1/2	15	17 1/2	20	22 1/2	25	27 1/2	30
1 1/2 x 4	4	5	6	7	8	7 1/2	8 1/3	9 1/6	10
1 1/2 x 6	6	7 1/2	9	10 1/2	12	11 1/4	12 1/2	13 3/4	15
1 1/2 x 8	8	10	12	14	16	15	20	22	24
1 1/2 x 10	10	12 1/2	15	17 1/2	20	18 3/4	25	27 1/2	30
1 1/2 x 12	12	15	18	21	24	22 1/2	30	33	36
2 x 4	5 1/3	6 2/3	8	9 1/3	10 1/3	12	13 1/3	14 2/3	16
2 x 6	8	10	12	14	16	18	20	22	24
2 x 8	10 2/3	13 1/3	16	18 2/3	21 1/3	24	26 2/3	29 1/3	32
2 x 10	13 1/3	16 2/3	20	23 1/3	26 2/3	30	33 1/3	36 2/3	40
2 x 12	16	20	24	28	32	36	40	44	48
3 x 6	12	15	18	21	24	27	30	33	36
3 x 8	16	20	24	28	32	36	40	44	48
3 x 10	20	25	30	35	40	45	50	55	60
3 x 12	24	30	36	42	48	54	60	66	72
4 x 4	10 2/3	13 1/3	16	18 2/3	21 1/3	24	26 2/3	29 1/3	32
4 x 6	16	20	24	28	32	36	40	44	48
4 x 8	21 1/3	26 2/3	32	37 1/3	42 2/3	48	53 1/3	58 2/3	64
4 x 10	26 2/3	33 1/3	40	46 2/3	53 1/3	60	66 2/3	73 1/3	80
4 x 12	32	40	48	56	64	72	80	88	96

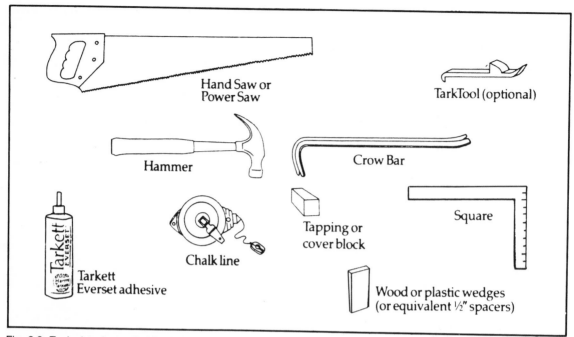

Fig. 2-2. Typical tools needed for hardwood flooring installation (courtesy Harris-Tarkett, Inc.).

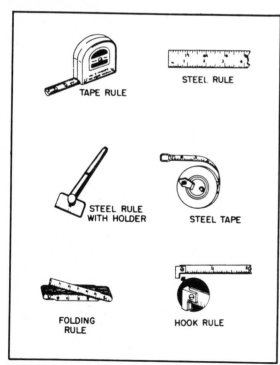

Fig. 2-3. Common types of rules and tapes.

tions are not, as a rule, numbered individually, however, since they are sufficiently far apart to be counted without difficulty. The opposite side is similarly divided into 32 and 64 spaces per inch. It is common practice to number every fourth division of these graduations for easier reading.

There are many variations of the common rule. Sometimes the graduations are on one side only. Sometimes a set of graduations are added across one end for measuring in narrow spaces. And sometimes only the first inch is divided into 64ths, with the remaining inches divided into 32nds and 16ths.

For measuring lengths greater than 18 inches, folding steel, wood, or aluminum rules can be used. They are usually 2 to 6 feet long. The folding rules cannot be relied on for extremely accurate measurements because a certain amount of play develops at the joints after they have been used for awhile.

Steel tapes are made from 6 to about 100 feet in length. In the shorter lengths, they are frequently made with a curved cross section so that they are flexible enough to roll up, but remain rigid when extended. Long, flat tapes require support over their full length when measuring, or the natural sag will cause an error in reading.

The flexible-rigid tapes are usually contained in metal cases into which they wind themselves when a button is pressed, or into which they can be easily wound. A hook is provided at one end to hook over the object being measured so one person can handle it without assistance. On some models, the outside of the case can be used as one end of the tape when inside dimensions are being measured.

Rules and tapes should be handled carefully and kept lightly oiled to prevent rust. Never allow the edges of measuring devices to become nicked by striking them with hard objects. They should preferably be kept in a wooden box when not in use.

To avoid kinking tapes, pull them straight out from their cases; do not bend them backward. With the windup type, always turn the crank clockwise. Turning it backward will kink or break the tape. With the spring-wind type, guide the tape by hand. If it is allowed to snap back, it may be kinked, twisted, or otherwise damaged.

USING A TAPE RULE

In Fig. 2-4, notice in the circle that the hook at the end of the perpendicular rule is attached so that it is free to move slightly. When an outside dimension is taken by hooking the end of the rule over the edge, the hook will locate the end of the rule even with the surface from which the measurement is being taken.

To measure an inside dimension using a tape rule, extend the rule between the surfaces as shown, take a reading at the point on the scale where the rule enters the case, and add 2 inches. The 2 inches are the width of the case. The total is the inside dimension being taken.

To measure an outside dimension using a tape rule, hook the rule over the edge of the stock. Pull the tape out until it projects far enough from the case to permit you to measure the required distance. The hook is designed so that it will locate the end of the rule at the surface from which the

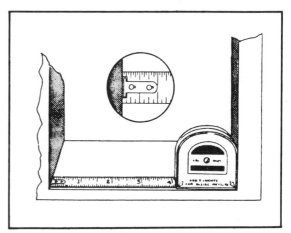

Fig. 2-4. Measuring inside dimensions with a tape rule.

measurement is being taken (Fig. 2-5). When you are taking a measurement of length, hold the tape parallel to the lengthwise edge. For measuring widths, the tape should be at right angles to the lengthwise edge. Read the dimension of the rule exactly at the edge of the piece being measured. In this case, it may be necessary to butt the end of the tape against another surface or to hold the rule at a starting point from which a measurement is to be taken.

USING A FRAMING SQUARE

Of all the layout tools in the woodworker's kit, the framing square is the most generally useful.

Fig. 2-5. Measuring outside dimensions with a tape rule.

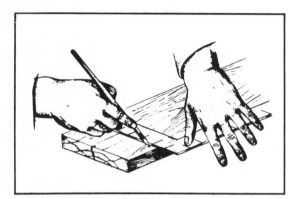

Fig. 2-6. Using a framing square.

In laying out 90-degree and 45-degree angles with a framing square, the lumber you work with should be squared on the ends. This will make it necessary to lay out a line at a 90-degree angle with respect to the edge of the board. This line should be as close to the end of the board as possible to avoid undue waste of material. When doing this job with a framing square, place the blade of the square along one edge of the board and mark along the outside edge of the tongue, as shown in Fig. 2-6.

USING A COMBINATION SQUARE

Another versatile tool for measuring, marking, and preparing hardwood floor strips or blocks for cutting is the combination square. To square a line on stock with a combination square, place the squaring head on the edge of the stock, as shown in Fig. 2-7, and draw the line along either edge of the blade. The line will be square with the edge of the stock against which the squaring head is held. The angle between the line and the edge will be 90 degrees.

To lay out a 45-degree angle on stock using a combination square, place the squaring head on the edge of the stock, as shown in Fig. 2-8, and draw the line along either edge of the blade. The line will be at 45 degrees with the edge of the stock against which the squaring head is held.

To test trueness of 90-degree angles with a combination square, hold the body of the square in contact with one surface of the 90-degree angle and bring the blade into contact with the other (Fig. 2-9). In making this test, have the square between yourself and a good source of light. If the angle is a true 90-degree angle, no light will be visible between the blade and the surface of the work.

There's one other useful tool that could be mentioned here: the chalk line (Fig. 2-10). A chalk line is a white, twisted mason's line consisting of a reel, line, and chalk. It is coated with chalk and stretched taut between points to be connected by a straight line, just off the surface. When snapped, the line makes a straight guideline. The chalk line can be very useful in making diagonal cuts across hardwood floor boards.

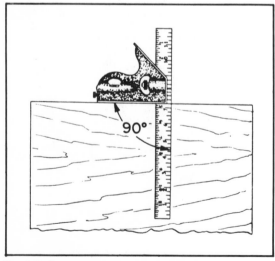

Fig. 2-7. Squaring stock with a combination square.

Fig. 2-8. Making 45-degree cuts with a combination square.

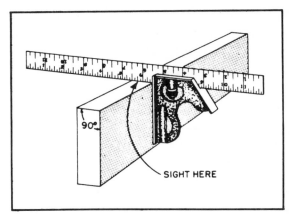

Fig. 2-9. Checking a 90-degree angle with a combination square.

HANDSAWS

Anyone working with wood uses a large variety of hand tools. Of these, woodcutting hand tools are often the most used—and abused.

The most common carpenter's handsaw (Fig. 2-11) consists of a steel blade with a handle at one end. The blade is narrower at the end opposite the handle, which is called the *point* or *toe!* The end of the blade nearest the handle is called the *heel*. One edge of the blade has teeth that act as two rows of cutters. When the saw is used, these teeth cut two parallel grooves close together. The chips (*sawdust*) are pushed out from between the grooves (*kerf*) by the beveled part of the teeth. The teeth are bent alternately to one side or the other to make the kerf wider than the thickness of the blade. This bending is called the *set* of the teeth.

The number of teeth per inch, the size and shape of the teeth, and the amount of set depend on how the saw is used and the type of material to be cut. Carpenter's handsaws are described by the number of points per inch. There is always one more point than there are teeth per inch. A number stamped near the handle gives the number of points of the saw.

Woodworking handsaws consist of ripsaws and crosscut saws designed for general cutting. *Ripsaws* are used for cutting with the grain, and *crosscut saws* are for cutting across the grain. The major difference between a ripsaw and a crosscut saw is the shape of the teeth. A ripsaw tooth has a square-faced, chisel cutting edge (Fig. 2-12). It does a good job of cutting with the grain (called *ripping*), but a poor job of cutting across the grain (called *crosscutting*). A crosscut saw tooth has a beveled, knifelike cutting edge, like the one in Fig. 2-13. It does a good job of cutting across the grain, but a poor job of cutting with the grain. Figure 2-14 illustrates a combination saw blade that offers the qualities of both the ripsaw and the crosscut saw.

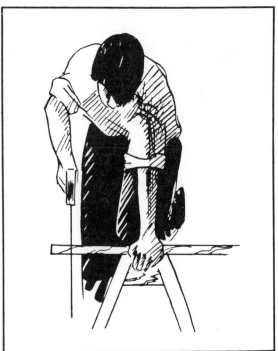

Fig. 2-11. Using a common carpenter's saw.

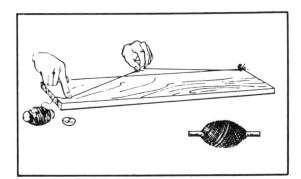

Fig. 2-10. Using a chalk and line.

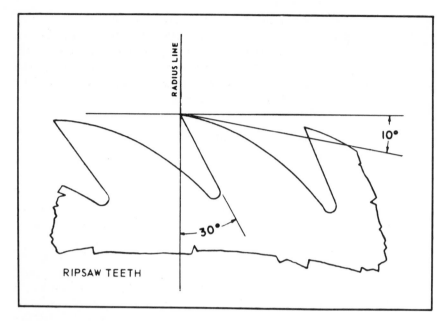

Fig. 2-12. Ripsaw teeth.

RADIUS LINE

10°

30°

RIPSAW TEETH

Special Handsaws

The more common types of saws used for special purposes are shown in Fig. 2-15. They can be useful as you notch hardwood floor strips or blocks around pipes or make critical cuts.

The *backsaw* is a crosscut saw designed for sawing a perfectly straight line across the face of a piece of stock. A heavy steel backing along the top of the blade keeps the blade perfectly straight. The *dovetail* saw is a special type of backsaw with a thin, narrow blade and a chisellike handle.

The *compass* saw is a long, narrow, tapering ripsaw designed for cutting out circular or other nonrectangular sections from within the margins of a board or panel. A hole is bored near the cutting line to start the saw. The *keyhole* saw is simply a finer, narrower compass saw. The *coping* saw is used to cut along the curved lines, as shown in Fig. 2-15.

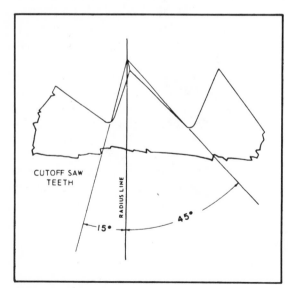

CUTOFF SAW TEETH

RADIUS LINE

15°   45°

Fig. 2-13. Cutoff saw teeth.

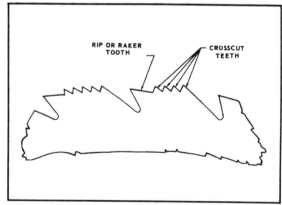

RIP OR RAKER TOOTH

CROSSCUT TEETH

Fig. 2-14. Combination saw teeth.

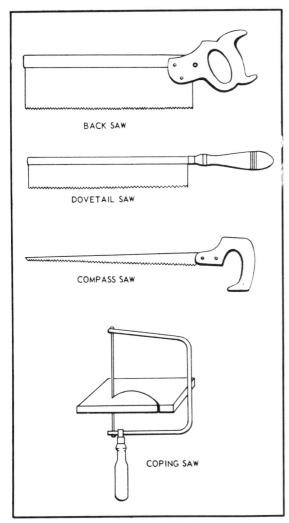

BACK SAW

DOVETAIL SAW

COMPASS SAW

COPING SAW

Fig. 2-15. Special saws for special jobs.

Caring for Handsaws

Some right and wrong methods of using and caring for a handsaw are shown in Fig. 2-16. A saw that is not being used should be hung up or stored in a toolbox. A toolbox designed for holding saws has notches that hold them on edge, teeth up. If you store saws loose in a toolbox, the saw teeth may become dulled or bent by contacting other tools.

Before using a saw, be sure there are no nails or other edge-destroying objects in the line of the cut. When sawing out a strip of waste, don't break out the strip by twisting the saw blade. Doing so

dulls the saw and may spring or break the blade.

Be sure that the saw will go through the full stroke without striking the floor or some other object. If the work cannot be raised high enough to obtain full clearance for the saw, you must carefully limit the length of each stroke.

SAWING AIDS

A *miter box* (Fig. 2-17) permits you to saw a piece of stock at a given angle without laying out a line. The figure shows a common type of wooden 45-degree miter box. Stock can be cut at 45 degrees by placing the saw in cuts M-S and L-F; or at 90 degrees by placing the saw in cuts A-B.

The *sawhorse* (Fig. 2-18) might be called the carpenter's portable workbench and scaffold. If you don't already have a good sawhorse, follow the illustration to make one. Not only is the sawhorse practical, but making one is a good exercise in using your handsaw and measuring tools and in gaining understanding of wood.

PORTABLE CIRCULAR SAW

The portable electric circular saw (Fig. 2-19) is one of the most popular tools for the do-it-yourselfer. It can be used for any purpose available to handsaws, while offering the advantage of speed. The size of a portable circular saw is designated by the maximum diameter of the blade in inches that it will support within its guard.

To make an accurate ripping cut, the ripping guide (Fig. 2-20A) is set a distance away from the saw equal to the width of the strip to be ripped off and placed against the edge of the piece as a guide for the saw. For cutting off, the ripping guide is turned upside down so that it will be out of the way.

Circular saws use circular blades. The types of blades are the same as those available on handsaws: ripping, crosscutting, power saws and combination.

SAFETY WITH POWER SAWS

All portable, power-driven saws should be equipped with guards that will automatically adjust themselves to the work so that none of the teeth

27

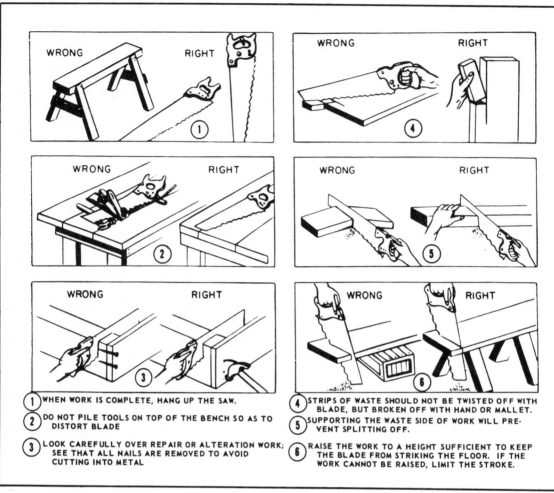

WRONG — RIGHT

① WHEN WORK IS COMPLETE, HANG UP THE SAW.

② DO NOT PILE TOOLS ON TOP OF THE BENCH SO AS TO DISTORT BLADE

③ LOOK CAREFULLY OVER REPAIR OR ALTERATION WORK; SEE THAT ALL NAILS ARE REMOVED TO AVOID CUTTING INTO METAL

④ STRIPS OF WASTE SHOULD NOT BE TWISTED OFF WITH BLADE, BUT BROKEN OFF WITH HAND OR MALLET.

⑤ SUPPORTING THE WASTE SIDE OF WORK WILL PREVENT SPLITTING OFF.

⑥ RAISE THE WORK TO A HEIGHT SUFFICIENT TO KEEP THE BLADE FROM STRIKING THE FLOOR. IF THE WORK CANNOT BE RAISED, LIMIT THE STROKE.

Fig. 2-16. Caring for handsaws.

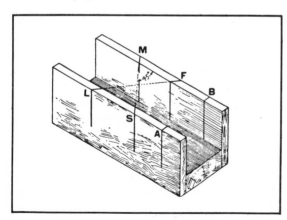

Fig. 2-17. Miter box.

protrude above the work. The guard over the blade should be adjusted so that it slides out of its recess and covers the blade to the depth of the teeth when the saw is lifted from the work. Goggles or face shields should be worn while using the saw and while cleaning up debris afterwards.

Saws are to be grasped with both hands and held firmly against the work. Care should be taken that the saw does not break away, thereby causing injury. The blade should be inspected at frequent intervals and always after it has locked, pinched, or burned. The electrical connection should be broken before this examination. The saw motor should not be overloaded by pushing too hard or cutting

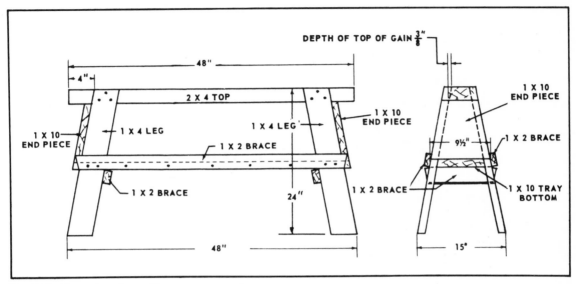

Fig. 2-18. Dimensions for making your own sawhorse.

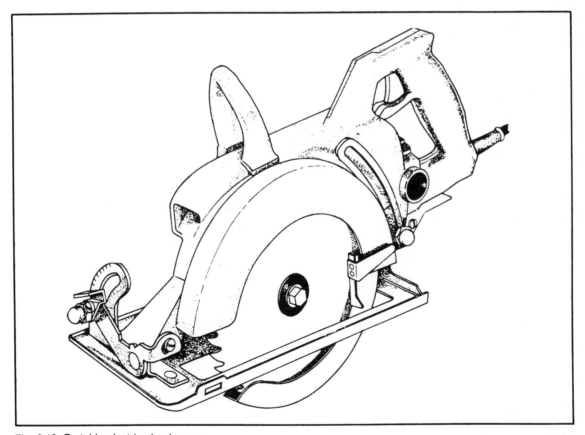

Fig. 2-19. Portable electric circular saw.

Fig. 2-20. Ripping with a portable circular saw, using a ripping guide (A).

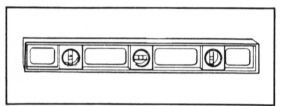

Fig. 2-21. Carpenter's level.

stock that is too heavy.

Before using the saw, material to be cut should be carefully examined and freed of nails or other metal substances. Cutting into or through knots should be avoided as much as possible.

The electric plug should be pulled before any adjustments or repairs are made to the saw, including changing the blade.

CARPENTER'S LEVEL

The carpenter's level (Fig. 2-21) determines the levelness of surface and sights level lines. It may be used directly on the surface or used with a straightedge (Fig. 2-22). Levelness is determined by bubbles suspended within glass tubes parallel to one or more surfaces of the level (Fig. 2-23).

To level a surface, such as the workbench in Fig. 2-24, set a carpenter's level on the bench top parallel to the front edge of the bench. Notice that the level may have as many as three or more pairs of glass vials. Regardless of the position of the level, always watch the bubble in the bottom vial of a horizontal pair. Shim or wedge up the end of the bench that will return that bubble to the center of

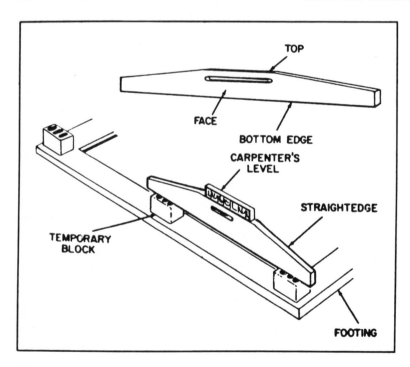

Fig. 2-22. Straightedge used with a carpenter's level.

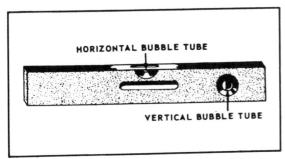

Fig. 2-23. Smaller carpenter's level with horizontal and vertical bubble tubes.

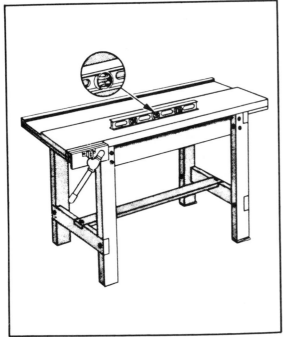

Fig. 2-24. Using the carpenter's level to check for a level working surface.

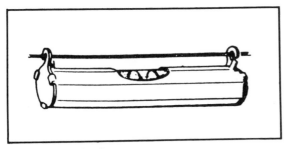

Fig. 2-25. Line level.

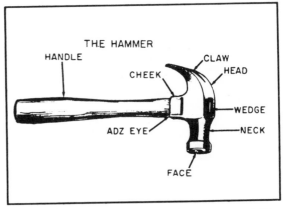

Fig. 2-26. Carpenter's curved-claw nail hammer components.

its vial. Recheck the first position of the level before securing the shims or wedges. These principles can easily be applied to the leveling of flooring material with a carpenter's level and shims.

The line level (Fig. 2-25) has a spirit bubble to show levelness as it is hung from a line. Placement halfway between the points to be leveled gives the greatest accuracy.

CLAW HAMMER

The carpenter's curved-claw nail hammer (Fig. 2-26) is a steel-headed, wooden-handled tool used for driving nails, wedges, and dowels. The *claw*, which is at one end of the head, is a two-pronged arch used to pull nails out of the wood. The other parts of the head are the *eye* and the *face*.

The face may be flat, in which case it is called a *plain face* (Fig. 2-27). The plain-faced hammer is easier for the beginner to learn to drive nails with. With this hammer, however, it is difficult to drive the nailhead flush with the surface of the work without leaving hammer marks on the surface.

The face of a hammer may also be slightly rounded, or convex, in which case it is called *bell-faced*. The bell-faced hammer is generally used in rough work. When handled by an expert, it can drive the nailhead flush with the surface of the work without damaging the surface.

To use the hammer, grasp the handle with the end flush with the lower edge of the palm (Fig. 2-28). Keep your wrist limber and relaxed. Grasp

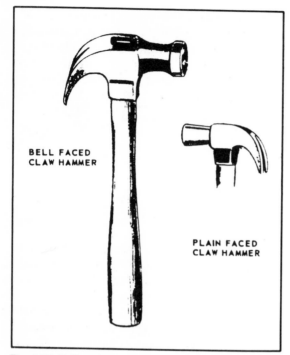

Fig. 2-27. Bell- and plane-faced claw hammers.

the nail with the thumb and forefinger of your other hand and place the point at the exact spot where it is to be driven. Unless the nail is to be purposely driven at an angle, it should be perpendicular to the surface of the work. Strike the nailhead squarely, keeping your hand level with the head of the nail. To drive, first rest the face of the hammer on the head of the nail, then raise the hammer slightly and give the nail a few light taps to start it and fix the aim. Then take your fingers away from the nail and drive the nail with firm blows with the center of the hammer face. The wrong way to drive a nail is shown in Fig. 2-29.

NAILING

Nailing two pieces of wood together is one of the most common tasks in carpentry and one that you'll be able to practice many times as you install your hardwood floors. If the joint doesn't hold or the wood splits, it is generally because the installer didn't observe a few basic rules for nailing.

First of all, if the joint is to hold properly, the nail must be long enough. A good rule to follow here is to select a nail three times the length of the thickness of the wood to be nailed. If the nail is too short, it cannot hold properly; if it is too long, the increased diameter may split the wood. There will be more information on selecting nails later in this chapter.

A few properly spaced nails will hold better than many nails put in at one point. Improper spacing or too many nails will split the wood and add no strength to the joint. A nail driven in at an angle, called *toenailing* (Fig. 2-30), will provide a stronger joint than if the nail is driven straight down.

When the end of the nail extends through the second piece of wood, it should be clinched or bent over. Although clinching the nail with the grain will give a smoother surface, clinching across the grain will give more strength.

The head of a finishing nail should be set below the surface of the wood (Fig. 2-31) with a nail set and the resulting hole filled with putty or plastic

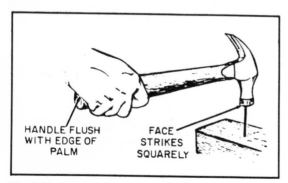

Fig. 2-28. Correct way to use a hammer.

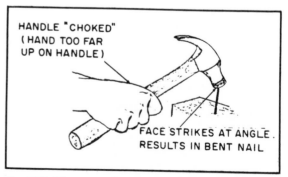

Fig. 2-29. Wrong way to use a hammer.

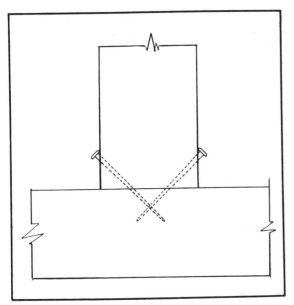

Fig. 2-30. Toenailing.

wood. Whenever a finished appearance is desired, drive the nail almost to the surface with the hammer and finish the job with a nail set (Fig. 2-32). This method will prevent your striking the wood with the face of the hammer and denting it.

When you are using a claw hammer to pull out nails, a block of wood should be inserted under the hammer to provide more leverage and to prevent the hammer from damaging the wood. If the nail has been clinched, it should be straightened before any attempt is made to remove it.

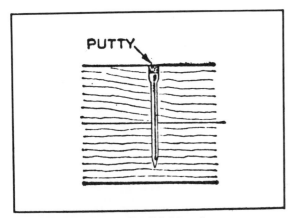

Fig. 2-31. Filling over a "set" finish nail.

Considering how inexpensive nails are, it is a waste of time and materials to try to straighten used nails for reuse. Once a nail has been used and bent back and forth, it has lost much of its holding power. Moreover, it is almost impossible to straighten a nail perfectly, and the usual result is the bending of the nail as it is driven in, which entails the removal of the nail and possible damage to the wood.

NAILS

Figure 2-33 shows the more common types of wire nails. The brad and the finish nail both have a deep countersink head that is designed to be set below the surface of the work. The casing nail has a flat countersink head, which may be driven flush and left that way or which may also be set. The other nails shown are all flat-headed nails.

The common nail is the one most widely used in general wood construction. Nails with large flat heads are used for nailing roofing paper, plasterboard, and similar thin or soft materials. Duplex or double-headed nails are used for nailing temporary

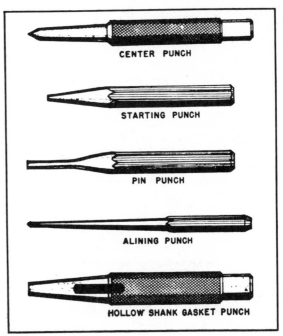

Fig. 2-32. Commonly used punches.

33

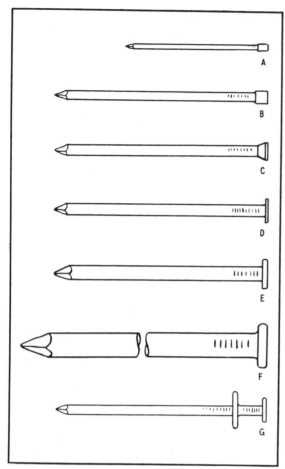

Fig. 2-33. Common types of wire nails: (A) brad, (B) finish nail, (C) casing nail, (D) box nail, (E) common nail, (F) spike (larger than 60d), (G) duplex head nail.

limited for the do-it-yourselfer, and the skill required for placement is high; so they are usually not used.

The lengths of the most commonly used nails are designated by the *penny* system. This system originated in England where the abbreviation for the word *penny* is the letter *d*. Thus, the expression "two-penny nail" is written "2d nail." The thickness of a nail increases with the penny size, and the number of nails per pound decreases. A box or casing nails of the same common nail penny size is thinner, and it takes more of them to weigh the same amount.

The penny sizes and corresponding length, thicknesses (in gauge sizes), and numbers per pound of the most commonly used nails are shown in Table 2-4. Relative sizing is shown in Fig. 2-35. Recommended nailing methods and sizes for specific home construction projects are given in Table 2-5.

The lengths of nails larger than 60d (called *spikes*) are designated in inches. Nails smaller than 2d are designated in fractions of an inch.

WOOD SCREWS

Wood screws (Fig. 2-36) are designated by the type of head and the material, such as flathead brass or roundhead steel. Most wood screws are made of either steel or brass, but there are copper and bronze wood screws as well. To distinguish the

structures, such as scaffolds, that will eventually be dismantled. A duplex nail has an upper and lower head. The nail is driven to the lower head; it can be easily drawn by setting the claw of a hammer under the upper head.

Besides nails with the usual type of shank (round), there are various special-purpose nails with shanks of other types. Nails with square, triangular, longitudinally grooved, and spirally grooved shanks have a much greater holding power than wire nails of the same size. One you may come across is the flooring nail (Fig. 2-34), driven diagonally through the tongue of the board into the subfloor. The availability of these nails, however, is somewhat

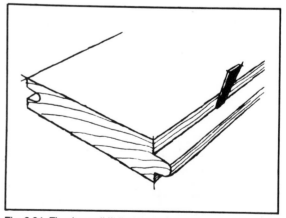

Fig. 2-34. Flooring nail "blind-nailed" into tongue-and-groove hardwood flooring strip.

Table 2-4. Sizes of Commonly Used Nails.

COMMON WIRE NAILS

Size	Length	Gage	Approx. No. to Lb.	Size	Length	Gage	Approx. No. to Lb.
2D	1 In.	No. 15	876	10D	3 In.	No. 9	69
3D	1-1/4	14	568	12D	3-1/4	9	63
4D	1-1/2	12-1/2	316	16D	3-1/2	8	49
5D	1-3/4	12-1/2	271	20D	4	6	31
6D	2	11-1/2	181	30D	4-1/2	5	24
7D	2-1/4	11-1/2	161	40D	5	4	18
8D	2-1/2	10-1/4	106	50D	5-1/2	3	14
9D	2-3/4	10-1/4	96	60D	6	2	11

FLOORING BRADS / FINISHING NAILS

Size	Length	Gage	Approx. No. to Lb.	Size	Length	Gage	Approx. No. to Lb.
6D	2 In.	No. 11	157	2D	1 In.	No. 16-1/2	1351
				3D	1-1/4	15-1/2	807
7D	2-1/4	11	139	4D	1-1/2	15	584
				5D	1-3/4	15	500
8D	2-1/2	10	99	6D	2	13	309
				7D	2-1/4	13	238
9D	2-3/4	10	90	8D	2-1/2	12-1/2	189
				9D	2-3/4	12-1/2	172
10D	3	9	69	10D	3	11-1/2	121
12D	3-1/4	8	54	12D	3-1/4	11-1/2	113
16D	3-1/2	7	43	16D	3-1/2	11	90
20D	4	6	31	20D	4	10	62

CASING NAILS / SMOOTH & BARBED BOX NAILS

Size	Length	Gage	Approx. No. to Lb.	Size	Length	Gage	Approx. No. to Lb.
2D	1 In.	No. 15-1/2	1010	2D	1 In.	No. 15-1/2	1010
3D	1-1/4	14-1/2	635	3D	1-1/4	14-1/2	635
4D	1-1/2	14	473	4D	1-1/2	14	473
5D	1-3/4	14	406	5D	1-3/4	14	406
6D	2	12-1/2	236	6D	2	12-1/2	236
7D	2-1/4	12-1/2	210	7D	2-1/4	12-1/2	210
8D	2-1/2	11-1/2	145	8D	2-1/2	11-1/2	145
9D	2-3/4	11-1/2	132	9D	2-3/4	11-1/2	132
10D	3	10-1/2	94	10D	3	10-1/2	94
12D	3-1/4	10-1/2	87	12D	3-1/4	10-1/2	88
16D	3-1/2	10	71	16D	3-1/2	10	71
20D	4	9	52	20D	4	9	52
30D	4-1/2	9	46	30D	4-1/2	9	46
				40D	5	8	35

ordinary type of head from the *Phillips* head, the former is called a *slotted* head. A *lag screw* is a heavy iron screw with a square bolt-type head. Lag screws are used mainly for fastening heavy timbers, but they can also be used for the installation of hardwood plank floors.

The size of an ordinary wood screw is designated by the length and the boy diameter (unthreaded part) of the screw (Fig. 2-37). Body diameters are designated by gauge numbers running from 0 (for about a 1/16-inch diameter) to 24 (for about a 3/8-inch diameter). Lengths range from 1/4 inch to 5 inches. The length and gauge numbers are printed on the box, as "1 1/4-9." This means a No. 9 gauge screw 1 1/4 inches long.

Note that for a nail, a large gauge number means a small nail; but for a screw, a large gauge number means a large screw. Tables 2-6 and 2-7 will guide you in the sizing of wood screws.

Figure 2-38 illustrates how screws can be

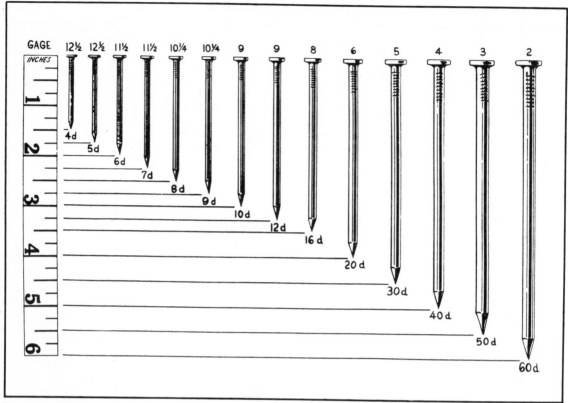

Fig. 2-35. Relative nail sizes.

countersunk through hardwood flooring and the subflooring. First, drill the body hole completely through the flooring. Then drill the starter hole, which is a little less than the diameter of the wood screw. Finally, if a flathead wood screw or ovalhead wood screw is to be used, countersink the body hole as shown.

Thus far in this chapter, you've learned a great deal about planning hardwood floors. Now you will see how a home is constructed from the ground up, with emphasis on how it is prepared for the installation of hardwood flooring.

THE FOUNDATION

One of the first essentials in house construction is to select the most desirable property site for its location. A lot in a smaller city or community presents few problems. The front setback of the house and side yard distances are either controlled by local regulations or governed by other houses in the neighborhood.

The *footing* is located at the base of the foundation (Fig. 2-39). It is made of concrete and is wider than the actual foundation so that the weight of the house will be distributed over a greater area. If the footing is not the right size for the weight of the house and the soil conditions, it will sink and the house will tend to settle.

The *foundation* is the masonry on top of the footing (Fig. 2-40) and supports the weight of the house. It also provides the walls for the basement, if planned. The foundation can be made of stone, cement, cinder blocks, poured concrete, or any other material that can sustain a considerable load.

The *sills* (Figs. 2-41 through 2-43) are the wood or steel beams around the top of the foundation. These beams are attached, and the house is built up from them.

Girders (Figs. 2-44 through 2-46) are the large beams running between opposite sills. They are used to provide additional support for the frame of the house, as well as carry the flooring.

The *floor joists* (Figs. 2-47 and 2-48) are the beams that run across the sills and provide a base for the flooring. Floor joists are generally made of 2-×-10-inch or 2-×-8-inch lumber, depending upon the distance they must run. They are placed broadside upright for greater strength, and in well-constructed homes, they are spaced 16 inches from center to center.

The *bridging* (Fig. 2-49) consists of small strips of 1-×-3-inch lumber, or a size near this, that are

Table 2-5. Recommended Nailing for Houses.

Joining	Nailing method	Nails		
		Number	Size	Placement
Header to joist	End-nail	3	16d	
Joist to sill or girder	Toenail	2–3	10d or 8d	
Header and stringer joist to sill	Toenail		10d	16 inches on center.
Bridging to joist	Toenail each end	2	8d	
Ledger strip to beam, 2 inches thick		3	16d	At each joist.
Subfloor, boards:				
1 by 6 inches and smaller		2	8d	To each joist.
1 by 8 inches		3	8d	To each joist.
Subfloor, plywood:				
At edges			8d	6 inches on center.
At intermediate joists			8d	8 inches on center.
Subfloor (2 by 6 inches, T&G) to joist or girder	*lind-nail* (casing) and face-nail.	2	16d	
Soleplate to stud, horizontal assembly	End-nail	2	16d	At each stud.
Top plate to stud	End-nail	2	16d	
Stud to soleplate	Toenail	4	8d	
Soleplate to joist or blocking	Face-nail		16d	16 inches on center.
Doubled studs	Face-nail, stagger		10d	16 inches on center.
End stud of intersecting wall to exterior wall stud	Face-nail		16d	16 inches on center.
Upper top plate to lower top plate	Face-nail		16d	16 inches on center.
Upper top plate, laps and intersections	Face-nail	2	16d	
Continous header, 2 pieces, each edge			12d	12 inches on center.
Ceiling joist to top wall plates	Toenail	3	8d	
Ceiling joist laps at partition	Face-nail	4	16d	
Rafter to top plate	Toenail	2	8d	
Rafter to ceiling joist	Face-nail	5	10d	
Rafter to valley or hip rafter	Toenail	3	10d	
Ridge board to rafter	End-nail	3	10d	
Rafter to rafter through ridge board	Toenail	4	8d	
	Edge-nail	1	10d	
Collar beam to rafter:				
2-inch member	Face-nail	2	12d	
1-inch member	Face-nail	3	8d	
1-inch diagonal let-in brace to each stud and plate (4 nails at top).		2	8d	
Built-up corner studs:				
Studs to blocking	Face-nail	2	10d	Each side.
Intersecting stud to corner studs	Face-nail		16d	12 inches on center.
Built-up girders and beams, 3 or more members	Face-nail		20d	32 inches on center, each side.
Wall sheathing:				
1 by 8 inches or less, horizontal	Face-nail	2	8d	At each stud.
1 by 6 inches or greater, diagonal	Face-nail	3	8d	At each stud.
Wall sheathing, vertically applied plywood:				
⅜ inch and less thick	Face-nail		6d	⎱6-inch edge.
½ inch and over thick	Face-nail		8d	⎰12-inch intermediate.
Wall sheathing, vertically applied fiberboard:				
½ inch thick	Face-nail			1½-inch roofing nail.[1]
25⁄32 inch thick	Face-nail			1¾-inch roofing nail.[1]
Roof sheathing, boards, 4-, 6-, 8-inch width	Face-nail	2	8d	At each rafter.
Roof sheathing plywood:				
⅜ inch and less thick	Face-nail		6d	⎱6-inch edge and 12-
½ inch and over thick	Face-nail		8d	⎰ inch intermediate.

[1] 3-inch edge and 6-inch intermediate.

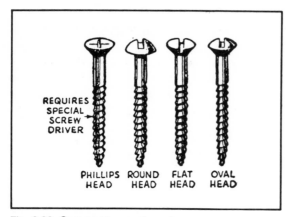

Fig. 2-36. Common types of wood screws.

nailed diagonally between the floor joists along the center of the span. The purpose of the bridging is to keep the joists perpendicular so that they will provide the maximum amount of support and to distribute the weight on the floor between several joists rather than one or two. Bridging can also be made out of strips of metal.

The subfloor (Fig. 2-50) is the underflooring to which the finish floor is nailed. The subfloor is nailed directly to the floor joists and runs either at a 45- or a 90-degree angle to them. The subfloor or rough floor not only furnishes a base for the finish floor, but also adds a degree of strength to

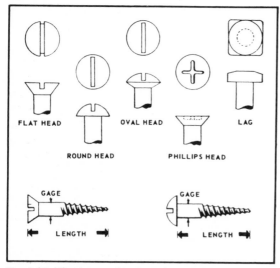

Fig. 2-37. Wood screw heads and components.

the frame of the house. Figures 2-51 and 2-52 show the actual installation of subflooring.

An important part of energy efficiency in the home is the installation of adequate insulation under the flooring. Figure 2-53 illustrates how batt insulation is placed under flooring. Such batt insulation is usually 48 inches long and 15 or 23 inches wide, depending on whether the joists are placed 16 or 24 inches apart.

WALL FRAMING

Wall framing (Fig. 2-54) is composed of regular studs, diagonal bracing, cripples, trimmers, headers, and fire blocks and is supported by the floor sole plate. The vertical members of the wall framing are the studs, which support the top plates and all the weight of the upper part of the building or everything above the top plate line. They provide the framework to which the wall sheathing is nailed on the outside and which supports the wall covering and insulation on the inside.

Walls and partitions classed as framed constructions are composed of structural elements called *studs*, that are usually closely spaced, slen-

Table 2-6. Screw Threads per Inch.

Diameter		Threads Per Inch			
No.	Inch	Decimal Equivalent	NC	NF	EF
0	----	.0600	---	80	---
1	----	.0730	64	72	---
2	----	.0860	56	64	---
3	----	.0990	48	56	---
4	----	.1120	40	48	---
5	----	.1250	40	44	---
6	----	.1380	32	40	---
8	----	.1640	32	36	---
10	----	.1900	24	32	40
12	----	.2160	24	28	---
---	1/4	.2500	20	28	36
---	5/16	.3125	18	24	32
---	3/8	.3750	16	24	32
---	7/16	.4375	14	20	28
---	1/2	.5000	13	20	28
---	9/16	.5625	12	18	24
---	5/8	.6250	11	18	24
---	3/4	.7500	10	16	20
---	7/8	.8750	9	14	20
---	1	1.0000	8	14	20

Table 2-7. Screw Sizes and Dimensions.

Length (in.)	Size numbers																					
	0	1	2	3	4	5	6	7	8	9	10	11	12	13	14	15	16	17	18	20	22	24
1/4	x	x	x	x																		
3/8	x	x	x	x	x	x	x	x	x	x												
1/2			x	x	x	x	x	x	x	x	x	x	x									
5/8		x	x	x	x	x	x	x	x	x	x	x	x		x							
3/4			x	x	x	x	x	x	x	x	x	x	x		x		x					
7/8			x	x	x	x	x	x	x	x	x	x	x		x		x					
1				x	x	x	x	x	x	x	x	x	x		x		x		x	x		
1 1/4					x	x	x	x	x	x	x	x	x		x		x		x	x		x
1 1/2					x	x	x	x	x	x	x	x	x		x		x		x	x		x
1 3/4						x	x	x	x	x	x	x	x		x		x		x	x		x
2						x	x	x	x	x	x	x	x		x		x		x	x		x
2 1/4						x	x	x	x	x	x	x	x		x		x		x	x		x
2 1/2						x	x	x	x	x	x	x	x		x		x		x	x		x
2 3/4							x	x	x	x	x	x	x		x		x		x	x		x
3							x	x	x	x	x	x	x		x		x		x	x		x
3 1/2									x	x	x	x	x		x		x		x	x		x
4									x	x	x	x	x		x		x		x	x		x
4 1/2												x			x		x		x	x		x
5												x			x		x		x	x		x
6															x		x		x	x		x
Threads per inch	32	28	26	24	22	20	18	16	15	14	13	12	11		10		9		8	8		7
Diameter of screw (in.)	.060	.073	.086	.099	.112	.125	.138	.151	.164	.177	.190	.203	.216		.242		.268		.294	.320		.372

der vertical members. They are arranged in a row with their ends bearing on a long horizontal member, called a *bottom plate* or *sill plate*, and their tops are capped with another plate, called a *top plate*. Double top plates are used in bearing walls and partitions. The bearing strength of stud walls is determined by the strength of the studs.

Figure 2-55 illustrates how an intersecting partition wall is tied into the exterior wall with double studs. Notice that these plates and walls all rest on the subfloor which, in turn, rests on joists.

Since you may be challenged by the installation of hardwood flooring on interior stairs, Fig. 2-56 shows a typical stairs installation and the many parts included. There are many different kinds of stairs, but all have two main parts in common: the *treads* people walk on, and the *stringers* (also called strings, horses, and carriages) which support the treads. A very simple type of stairway, consisting only of stringers and treads, is shown in the left-hand view of the figure. Treads of the type shown here are called *plank* treads. This simple type of stairway is called a *cleat* stairway because of the cleats attached to the stringers to support the treads.

A more finished type of stairway has the treads mounted on two or more sawtooth-edged stringers

Fig. 2-38. Countersinking holes for screws attaching hardwood flooring.

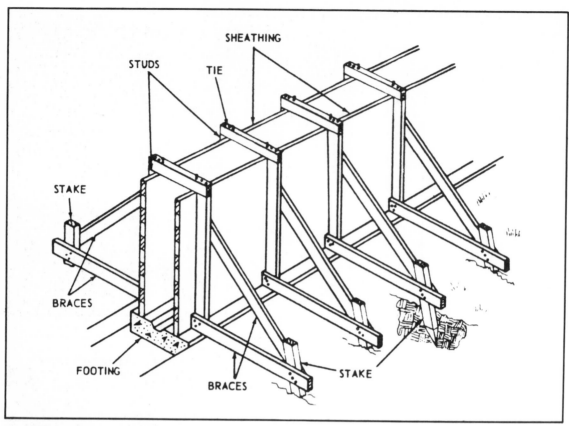

Fig. 2-39. Foundation and footing.

Fig. 2-40. Typical home's foundation.

40

Fig. 2-41. Foundation prior to sill installation.

Fig. 2-42. Close up of sill bolts imbedded in foundation.

Fig. 2-43. Sill plate installed with sill bolts.

Fig. 2-44. Flooring girders are installed on foundation piers.

Fig. 2-45. Girders fit into notches in the perimeter foundation.

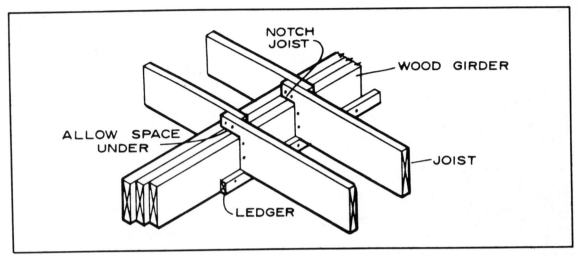

Fig. 2-46. Typical girder and intertied joists.

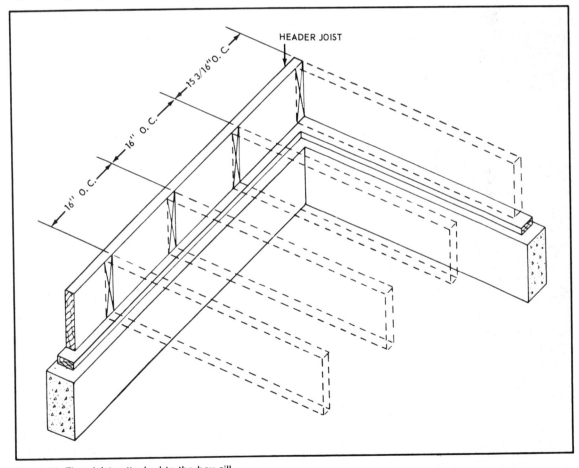

Fig. 2-47. Floor joists attached to the box sill.

43

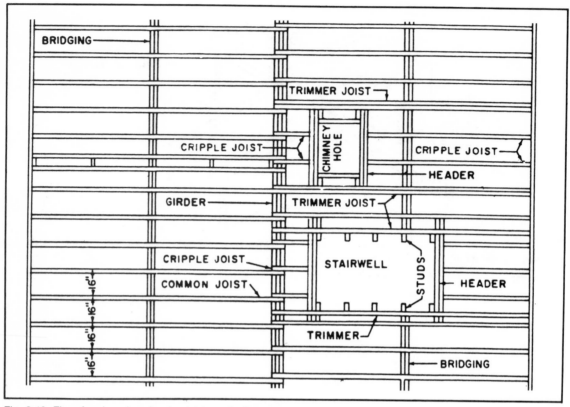

Fig. 2-48. Floor framing plan showing joists and other components.

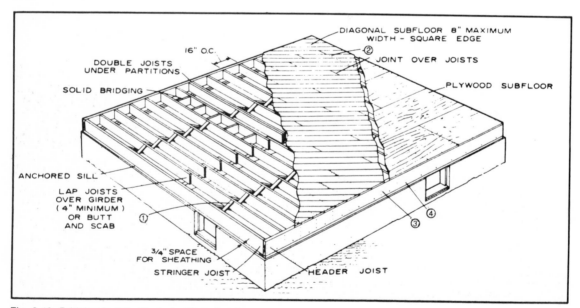

Fig. 2-49. Bridging installed in typical floor framing.

44

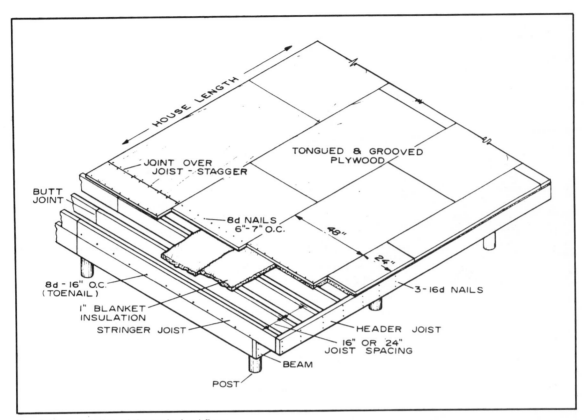

Fig. 2-50. Installation of the typical subfloor.

Fig. 2-51. Actual installation of subfloor.

Fig. 2-52. Close-up of subfloor installation.

and includes *risers*, as shown in the right-hand view of Fig. 2-56. The stringers shown here are cut out of solid pieces of dimensional lumber (usually 2-×-12) and are therefore called *cutout* or *sawed* stringers.

FLOORING

Figure 2-57 illustrates how the finish flooring is added on to the house construction. Figure 2-58 illustrates how square-edged flooring is nailed and installed over subflooring. First, building paper is installed on top of the subflooring. Then the finish flooring is installed at right angles and face-nailed.

Figure 2-59 shows the steps for installing tongue-and-groove finish flooring over a subfloor. Again, building paper can be applied between the subfloor and the finish floor in order to retard moisture from coming up through the subflooring and damaging the flooring.

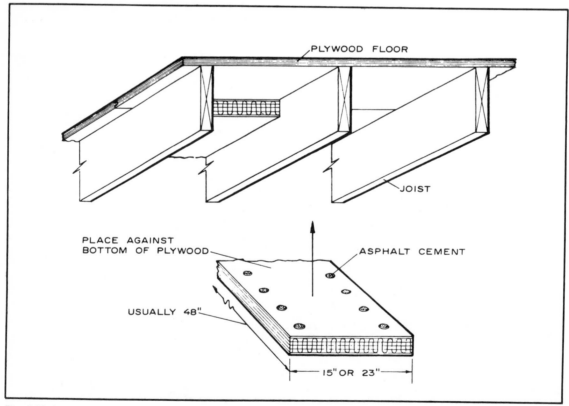

Fig. 2-53. Installing batt insulation under subfloor.

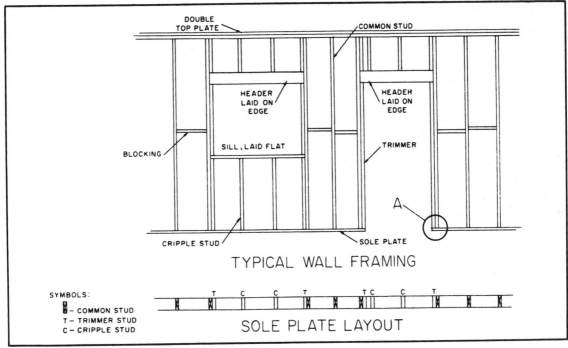

Fig. 2-54. Wall framing components including double-studded door frame (A).

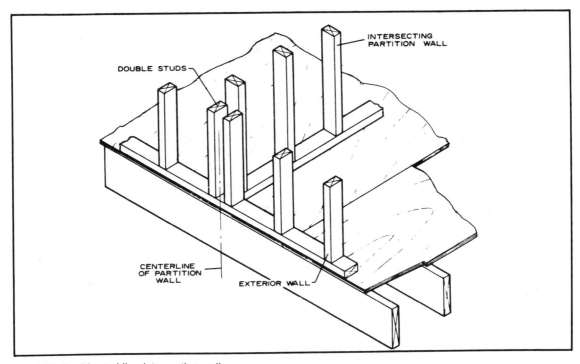

Fig. 2-55. Double-studding intersecting walls.

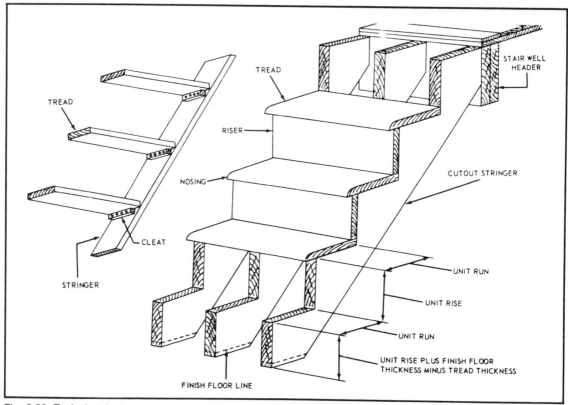

Fig. 2-56. Typical stairs installation.

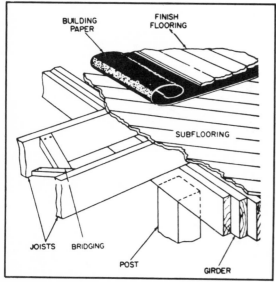

Fig. 2-57. Installation of finish flooring over joists, bridging, girder, and subflooring.

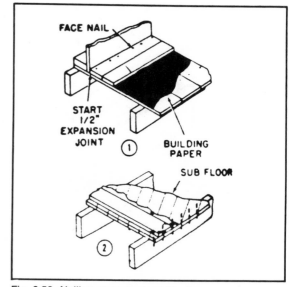

Fig. 2-58. Nailing square-edged flooring.

48

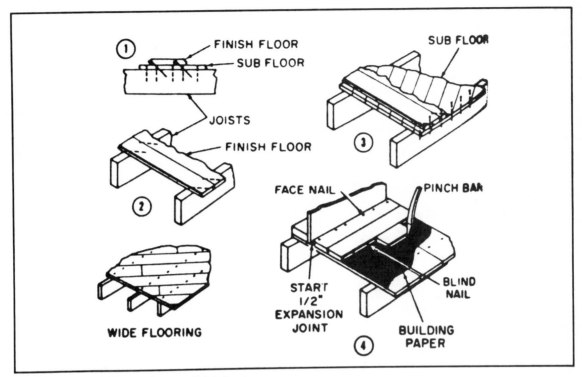

Fig. 2-59. Nailing tongue-and-groove finish flooring.

In the next chapter you'll learn the specific steps for the installation of hardwood strip and block flooring in a variety of situations and using numerous types of flooring available to the consumer. Remember that the planning of your hardwood flooring job is vital to its success.

Chapter 3

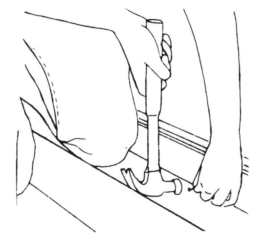

Installing Hardwood Floors

THE INSTALLAION OF HARDWOOD FLOORING is, in one way, complex. The variety of types, sizes, and application systems is so wide that there are literally dozens of ways of installing hardwood floors. Once you've selected the type of flooring, however, the installation becomes much easier. Each type of hardwood flooring has its own installation methods, and manufacturers and dealers are usually helpful in offering instructions.

The purpose of this chapter is to guide you through the installation of strip, block, and other types of hardwood flooring so that you can do a professional looking job—even if you've never worked with wood before. More than 50 illustrations will show you exactly how it's done.

INSTALLATION METHODS

Figure 3-1 illustrates the installation of wood strip flooring over a concrete slab floor. Since this type of floor tends to be cold and plain, many people install wood flooring over the top. First, be sure that there is sufficient insulation and a vapor bar-

rier between the slab and the ground, or at least the slab and the flooring. The anchored sleepers in the illustration are intended to allow space between the cold slab and the hardwood floor. Insulation can be installed here, if you desire.

Another way of installing strip flooring over a concrete slab is offered in Fig. 3-2. This method allows a greater air pocket between the concrete and the flooring with a waterproof coating and a vapor barrier to retard moisture—wood flooring's worst enemy. Nailing must be more precise in this method to fully anchor the flooring to the sleepers. This can be accomplished by using a chalk line to mark the nailing line or by careful nailing.

In both cases, be sure that treated lumber is used for the sleepers to ensure that the lumber won't rot and eventually attack your flooring from below.

Figures 3-3 and 3-4 illustrate the installation of flooring over a concrete slab at and below finish grade. I'll cover the installation of wood flooring over a concrete slab in more detail in the coming pages.

In most cases, strip and block flooring will be

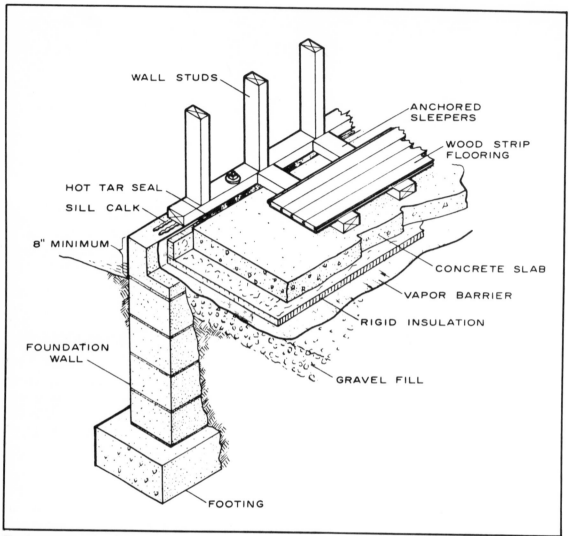

Fig. 3-1. Wood strip flooring over a concrete slab.

installed over wood subflooring, as discussed in Chapter 2. The typical wood subflooring system (Fig. 3-5) includes the floor joists, subflooring, building paper, and flooring. The flooring should be installed at a 45- or 90-degree angle (Fig. 3-6) to the subflooring.

HANDLING AND STORAGE

Most hardwood flooring has been kiln-dried to the proper moisture content. To maintain the mois-ture level, don't truck or unload it in rain, snow, or excessively humid conditions. Cover it with a tar-paulin or vinyl if the atmosphere in foggy or damp.

Check the job site before delivery. Be sure the flooring will not be exposed to high humidity or moisture. Surface drainage should direct water away from the building. Basements and crawl spaces must be dry and well-ventilated. In joist construc-tion with no basement, provision for outside cross ventilation must be provided through vents or other openings in the foundation walls. Total area of these

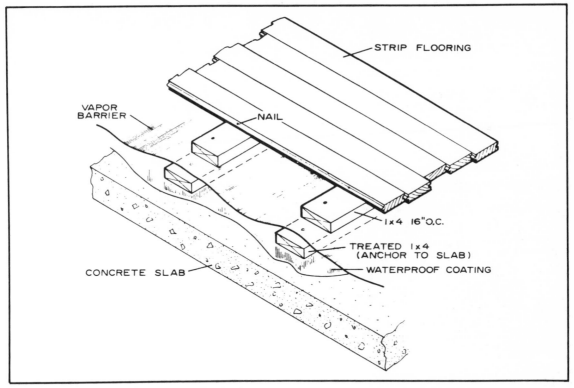

Fig. 3-2. Alternate installation over a concrete slab.

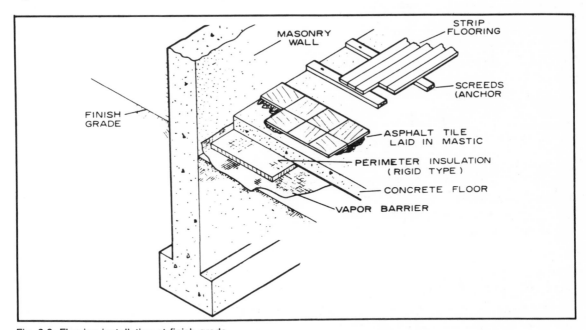

Fig. 3-3. Flooring installation at finish grade.

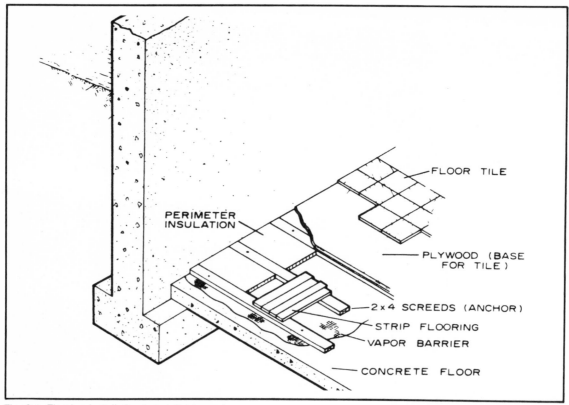

Fig. 3-4. Flooring installation below finish grade.

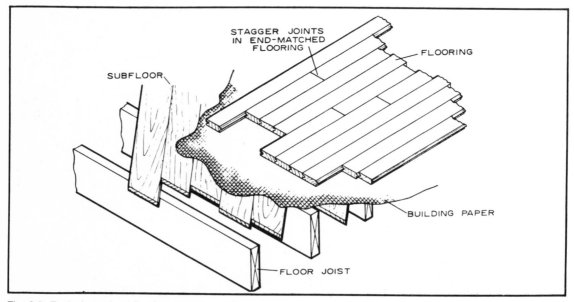

Fig. 3-5. Typical wood subflooring system.

Parquet flooring bordered by strip flooring (courtesy Jennison-Wright Corporation).

Hardwood strip flooring (courtesy Jennison-Wright Corporation).

Parquet flooring (courtesy Jennison-Wright Corporation).

Hardwood strip flooring set in a herringbone pattern (courtesy Jennison-Wright Corporation).

Fig. 3-6. Installing strip flooring at a 90-degree angle to subflooring (courtesy Dixon Lumber Company, Inc.).

openings should equal 1 1/2 percent of the first floor area. A ground cover of 4- or 6-mil polyethylene film is essential as a moisture barrier.

Flooring installed over a heating furnace or uninsulated ducts may develop cracks unless protection from the heat is provided. Use a double layer of 15-pound or a single layer of 30-pound asphalt felt or building paper, or 1/2-inch standard insulation board between joists under the flooring of these areas. Insulation used over a heating plant should be nonflammable.

The building should be closed in, with outside windows and doors in place in a newly constructed home. All concrete, plaster, and other masonry should be thoroughly dry before flooring is delivered to the job. In warm months the building must be well-ventilated; in winter, a temperature of 65 to 70 degrees Fahrenheit (no higher) should be maintained at least 5 days before the flooring is delivered for best results.

When job conditions are satisfactory, have the flooring delivered, broken up into small lots, and stored in the rooms where it will be installed. Allow 2 to 3 days for the flooring to become acclimated to the job site. Such protection from heat, cold, and moisture extremes may seem wasteful, but they will pay off in the installation and maintenance of your hardwood floors.

INSTALLATION OVER CONCRETE SLABS

All types of hardwood flooring can be installed successfully over a concrete slab. The slab must be constructed properly, however, and these instructions followed precisely.

Watch out for water. New concrete is heavy with moisture, an inherent enemy of wood. Proper on-grade slab construction requires a vapor barrier between the gravel fill and the slab. While this barrier prevents moisture entry through the slab, it also retards curing of the slab. So test for dryness, even if the slab has been in place over 2 years. To guide you, here are a few methods that professional builders and hardwood flooring installers use to test concrete for moisture.

The Rubber Mat Test. Lay a flat, noncorrugated rubber mat on the slab, place a weight on top to prevent moisture from escaping, and allow the mat to remain overnight. If there is "trapped" moisture in the concrete, the covered area will show water marks when the mat is removed. This test is only useful if the slab surface is light in color originally.

The Polyethylene Film Test. Tape a 1-foot square of heavy clear polyethylene film to the slab, sealing all edges with plastic packaging tape. If, after 24 hours, there is no "clouding," or drops of moisture on the underside of the film, the slab can be considered dry enough to install wood floors.

The Calcium Chloride Test. Place a 1/4 teaspoon of dry (anhydrous) calcium chloride crystals (available at drug stores) inside a 3-inch diameter putty ring on the slab. Cover with a glass so that the crystals are totally sealed off from the air. If the crystals dissolve within 12 hours, the slab is too wet for a hardwood flooring installation.

Keep in mind that the test should be made in several areas of each room on both old and new slabs. The remedy for a moist slab is to wait until it dries naturally, or accelerate drying with heat and ventilation.

Vapor Barrier

Start with a good vapor barrier. To be absolutely certain moisture doesn't reach the finished floor, a vapor barrier must be used on top of the slab. Its placement depends on the type of nailing surface and/or the type of wood flooring used. Prepare the slab by sweeping it clean. The slab must be sound, level, and free from grease, oil stains, and dust. Level out any high spots and fill low spots.

Plywood-on-Slab Method

This system uses 3/4-inch or thicker exterior plywood as the subfloor nailing base over the concrete slab. Refer to Fig. 3-7.

Roll out 4-mil or heavier polyethylene film over the entire slab, overlapping edges 4 to 6 inches and allowing enough to extend under the baseboard on all sides. It does not require imbedding in mastic.

Lay plywood panels out loose over the entire floor. Cut the first sheet of every run so end joints will be staggered 4 feet. Level 1/2-inch space at all wall lines and 1/4 about a 1/8-inch space.

Fasten the plywood with a powder-actuated concrete nailer or hammer-driven concrete nails. Use a minimum of nine nails per panel, starting at the

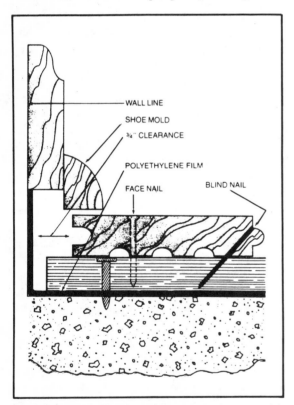

Fig. 3-7. Installing plywood subfloor over a concrete slab (courtesy National Oak Flooring Manufacturers Association).

center of the panel and working toward the edges to be sure of flattening out the plywood and holding it securely.

An alternate method is to cut the plywood into 4-×-4-foot squares, score the back, and lay on mastic adhesive. This method, however, will require use of either of two types of moisture barriers laid in mastic which will be described later in this chapter.

Screeds Method

The Screeds method of installing strip flooring on a concrete slab is shown in Fig. 3-8. This method uses flat, dry 2-×-4-inch screeds, or sleepers, of random lengths from 18 to 48 inches. They must be preservative-treated with a product other than creosote, which might bleed through and stain the finish floor.

Sweep the slab clean, prime with an asphalt primer, and allow to dry. Apply hot (poured) asphalt mastic and imbed the screeds, 12 to 16 inches on centers, at right angles to the direction of the finished floor. Stagger joints and lap ends at least 4 inches. Leave a 3/4-inch space between the ends of the screeds and the walls.

Over the screeds spread a vapor barrier of 4- or 6-mil polyethylene film with edges lapped 6 inches or more. It is not necessary to seal the edges or to affix the film with mastic, but avoid bunching or puncturing it, especially between screeds. The finish flooring will be nailed through the film to the screeds.

Some installers prefer to use a two-membrane asphalt felt or building paper vapor barrier, which will be explained later. The screeds are laid in rivers of mastic on the asphalt felt or building paper. In this system, the polyethylene film over the screeds is recommended for the extra moisture protection provided at nominal cost.

The screeds method alone—that is, without a subfloor and spaced 12 inches on center—is satisfactory for all strip flooring and plank flooring to 4-inch width. Plank flooring wider than 4 inches requires either the plywood-on-slab subfloor outlined earlier or screeds plus a wood subfloor to provide an adequate nailing surface. The subfloor may be 5/8-inch or thicker plywood or 3/4-inch boards.

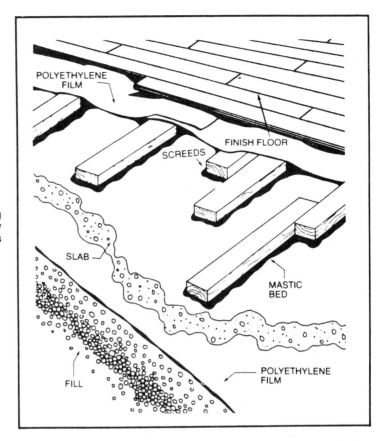

Fig. 3-8. Screeds method of installing flooring over concrete slab (courtesy National Oak Flooring Manufacturers Association).

Installing Block Flooring Over a Slab

Parquet, block, herringbone, and similar floors are normally laid in asphalt mastic and thus don't require a nailing surface on top of the slab. The need for a good moisture barrier is most important, however, and it can be achieved by either of the following methods.

Polyethylene Method. Prime the slab with an asphalt primer and allow to dry. Apply cold-type, cutback asphalt mastic with a straightedge trowel to the entire slab surface. Allow it to dry 30 minutes. Unroll 4-mil polyethylene film over the slab, covering the entire area and lapping the edges 4 inches. "Walk in" the film, stepping on every square inch of the floor to ensure proper adhesion. Small bubbles are of no concern.

Two-Membrane Asphalt Felt or Building Paper Method. Prime and apply mastic with a notched trowel (Fig. 3-9) at the rate of 40 square feet

per gallon. Let set 2 hours. Roll out 15-pound asphalt felt or building paper, lapping the edges 4 inches. Butt the ends. Apply another coating of mastic with the notched trowel and roll out a second layer

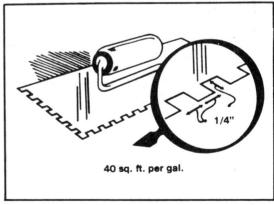

40 sq. ft. per gal.

Fig. 3-9. Notched trowel for applying mastic (courtesy Pennwood Products Co.).

of asphalt felt or building paper. Lay both layers in the same direction, but stagger the overlaps to achieve a more even thickness.

The finish floor will be laid in mastic on the vapor barrier. This method applies only to tongue-and-groove parquet flooring where tongues and grooves are engaged. It doesn't apply to slat-type or finger-block parquet.

INSTALLATION OVER WOOD JOISTS

Let's move on to the installation of hardwood flooring over wood joists. Use exterior plywood or boards of No. 1 or No. 2 common pine or other soft-wood suitable for subfloors over wood joists. If you use plywood, it must be at least 1/2 inch thick, preferably thicker. Lay panels with the grain of the faces at right angles to the joists and nail every 6 inches along each joist. Use appropriate nails for plywood thickness. Leave a 1/8-inch space between panels.

For a board subfloor (Fig. 3-10) use only flat, dry 3/4-inch dressed square-edge boards no wider than 6 inches. Lay them diagonally across the joists with 1/4-inch space between boards to allow for expansion. Don't use tongue-and-groove boards. Nail to every bearing point with two 8d or 10d common nails. All butt joints must rest on bearings.

Mark the location of joists so that flooring can be nailed into them. If subfloor boards are used over sleepers or screeds, allow 1/2-inch space between boards.

Good nailing is important. It keeps the board rigid, preventing creeping sometimes caused by shrinkage in subfloor lumber. Without adequate

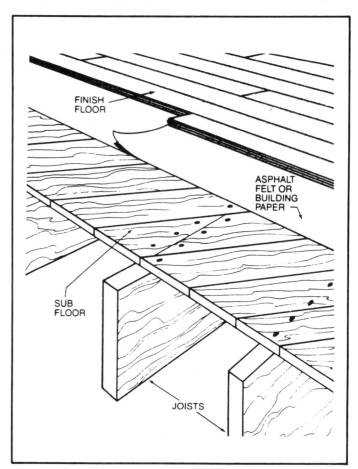

Fig. 3-10. Flooring over a board subfloor-ing (courtesy National Oak Flooring Manufacturers Association).

subfloor nailing, it is impossible to obtain solid, non-squeaking floors.

LAYING OUT THE FINISH FLOOR

Here are instructions for applying strip flooring laid on plywood-on-slab, on screeds, and on plywood or board subfloors.

When a plywood or board subfloor is used, start by renailing any loose areas and sweeping the subfloor clean. Then cover it with a good grade of 15-pound asphalt felt or building paper, lapping 4 inches at the seams to help keep out dust, retard moisture from below, and help prevent squeaks in dry seasons.

For the best appearance, lay the flooring in the direction of the longest dimension of the room or building—across or at right angles to the joists, as shown in Fig. 3-10. If a hallway parallels the long dimension of the room, begin the flooring by snapping a chalk line through the center of the hall and work from there into the room. Use a slip-tongue to reverse direction when you complete the hall later.

Location and straight alignment of the first course is important. Refer to Figs. 3-11 through 3-13. Place a strip of flooring 3/4 inch from the starter

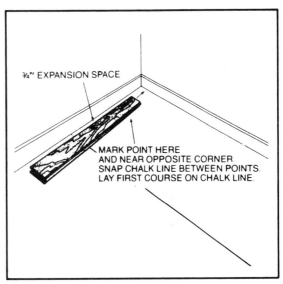

Fig. 3-11. Starter line for nailing first strip (courtesy National Oak Flooring Manufacturers Association).

wall (or leave as much space as will be covered by base and shoe mold), groove side toward the wall. Mark a point on the subfloor at the edge of the flooring tongue. Do this near both corners of the room, then snap a chalk line between the two points. Nail the first strip with its tongue on this line. The gap

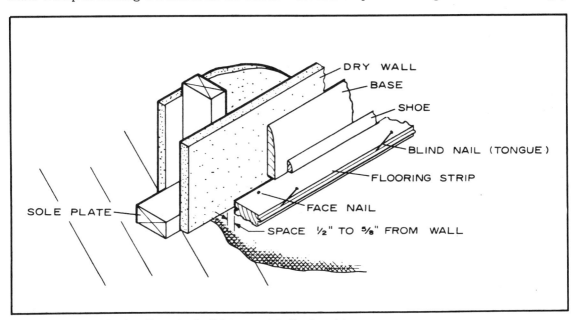

Fig. 3-12. How initial strip will eventually fit into complete flooring picture.

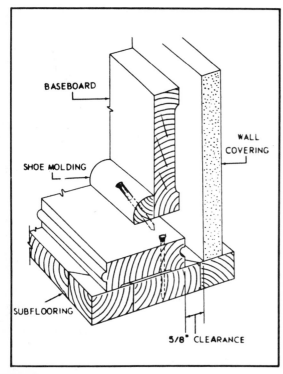

Fig. 3-13. Cross section of initial strip and molding.

BASEBOARD

WALL COVERING

SHOE MOLDING

SUBFLOORING

5/8" CLEARANCE

between the strip and the wall is needed for expansion space and will be hidden by the shoe mold.

If you're working with screeds on slab, you won't be able to snap a satisfactory chalk line on the loose polyethylene film laid over the screeds. Make the same measurements and stretch a line between nails at the wall edges. Remove the line after you get the starter board in place.

Lay the first strip along the starting chalk line, tongue out, and drive 8d finish nails at one end of the board near the grooved edge. Drive additional nails at each joist or screed and at midpoints between joists, keeping the starter strip aligned with the chalk line. Predrilled nail holes will prevent splits. The nailheads will be covered by the shoe molding. Nail additional boards in the same way to complete the first course.

Refer to Figs. 3-14 through 3-18. They illustrate the steps for matching, inserting, nailing, and setting nails when installing tongue-and-groove hardwood flooring.

Next, lay out seven or eight loose rows of flooring end to end in a staggered pattern with end joints at least 6 inches apart. Find or cut pieces to fit

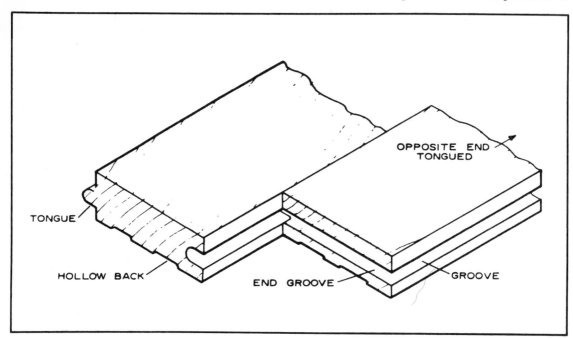

TONGUE

HOLLOW BACK

END GROOVE

GROOVE

OPPOSITE END TONGUED

Fig. 3-14. Matching tongue-and-groove flooring.

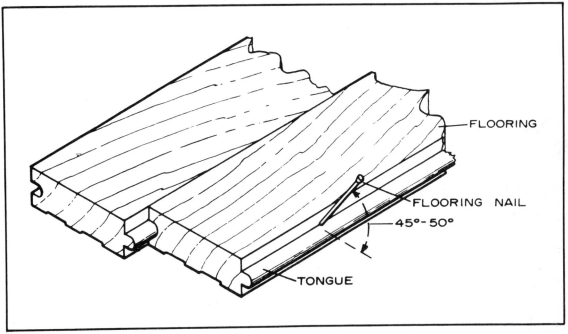

Fig. 3-15. Nailing tongue-and-groove flooring.

within 1/2 inch of the end wall. Watch your pattern for even distribution of long and short pieces and to avoid clusters of short boards.

Fit each board snug, groove to tongue, and blind-nail through the tongue according to Table 3-1. Refer to Figs. 3-19 and 3-20 for methods of wedging floor boards tight.

After the second or third course is in place, you can change from a hammer to a power nailer (Fig. 3-21), which is easier to use, does a much better job, and doesn't require countersinking. The power

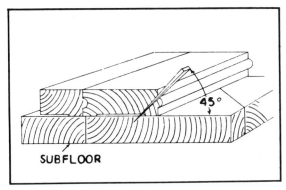

Fig. 3-16. Toenailing a flooring nail into the subfloor.

nailer drives a special barbed fastener, fed into the machine like staples, through the tongue of the floor at the proper angle. Power nailers can be rented through many rental yards or hardwood flooring suppliers.

When using the power nailer to fasten 3/4-inch strip or plank flooring to plywood laid on a slab, be sure to use a 1 3/4-inch cleat. The usual 2-inch cleat may come out the back of the plywood and prevent nails from countersinking properly. In all other applications, the 2-inch cleat is preferred.

Continue across the room, ending up on the far wall with the same 3/4-inch space allowed on the beginning wall. It may be necessary to rip a strip to fit.

Avoid nailing into a subfloor joint. If the subfloor joint is at right angles to the finish floor, don't let ends of the finish floor meet over it.

When nailing direct to screeds (no solid subfloor), nail at all screed intersections and to both screeds where a strip passes over a lapped screed joint. Since flooring ends are tongued-and-grooved, all end joints do not need to meet over screeds. End joints of adjacent strips, however, should not break

61

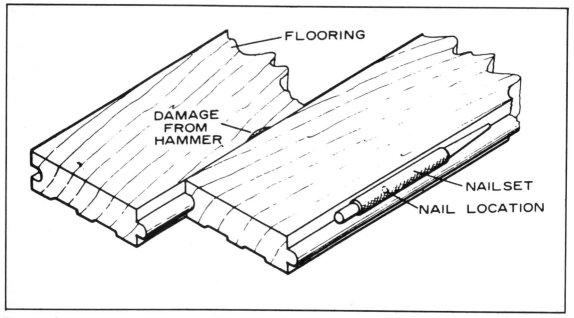

Fig. 3-17. Setting the nail to minimize damage to the hardwood flooring strip.

over the same void between screeds.

Some long boards may have horizontal bends, or *sweeps* resulting from a change in moisture content. A simple lever device (Fig. 3-22) can be made on the job to force such boards into position, as well as pull up several courses. Once the entire flooring is in place, nail the shoe molding to the baseboard, not to the flooring.

PLANK FLOORING

Plank flooring is normally made in 3- to 8-inch widths and may have countersunk holes for securing the planks with wood screws (Fig. 3-23). These holes are then filled with wood plugs, which are supplied with the flooring in many cases. Plank is installed in the same manner as strip flooring,

Fig. 3-18. Cross section of toenailed flooring strips.

Table 3-1. Nailing Schedule (Courtesy National Oak Flooring Manufacturers Association).

Flooring Nominal Size, Inches	Size of Fasteners	Spacing of Fasteners
¾ x 1½	2" machine driven fasteners; 7d or 8d screw or cut nail.	10"-12"* apart
¾ x 2¼	"	
¾ x 3¼	"	
¾ x 3" to 8" plank	"	8" apart into and between joists.
Following flooring must be laid on a subfloor.		
½ x 1½ ½ x 2	1½" machine driven fastener; 5d screw, cut steel or wire casing nail.	10" apart
⅜ x 1½ ⅜ x 2	1¼" machine driven fastener, or 4d bright wire casing nail.	8" apart
Square-edge flooring as follows, face-nailed — through top face		
5/16 x 1½ 5/16 x 2	1", 15-gauge fully barbed flooring brad. 2 nails every 7 inches.	
5/16 x 1⅓	1", 15-gauge fully barbed flooring brad. 1 nail every 5 inches on alternate sides of strip.	

*If subfloor is ½ inch plywood, fasten into each joist, with additional fastening between joists.

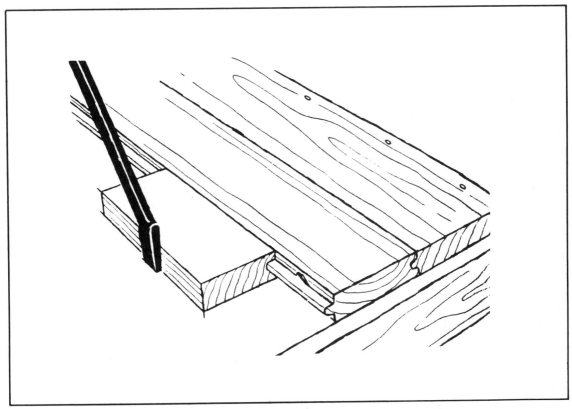

Fig. 3-19. Using a pinch bar and wedge to tighten flooring strips.

Fig. 3-20. Another method of keeping boards together, using a wood wedge.

Fig. 3-21. Using a power nailer to install strip flooring. Inset illustrates how barbed fastener is installed (courtesy National Oak Flooring Manufacturers Association).

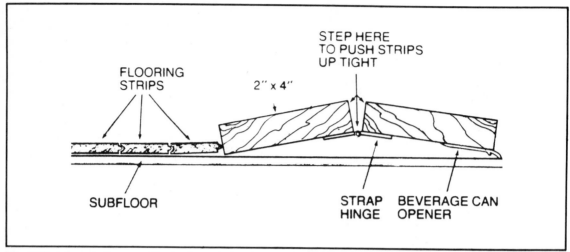

Fig. 3-22. Job-made lever used for forcing strip flooring boards up tight (courtesy National Oak Flooring Manufacturers Association).

alternating courses by widths. Start with the narrowest boards, then the next width, etc., and repeat the pattern.

Manufacturers' instructions for fastening the flooring vary and should be followed. The general

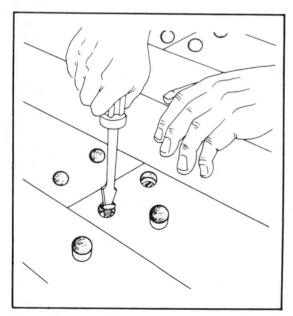

Fig. 3-23. Installing wood screws in countersunk holes drilled in plank flooring. Plugs are then inserted over screws (courtesy National Oak Flooring Manufacturers Association).

practice is to blind-nail through the tongue as with conventional strip flooring. Then countersink one or more (depending on the width of the plank) No. 9 or No. 12 screws at each end of each plank and at intervals along the plank to hold it securely. Cover the screws with wood plugs glued into the holes. Take care not to use too many screws because with the plugs in place, they tend to give the flooring a polka-dot appearance.

Be sure the screws are the right length. Use 1 inch if the floor is laid over 3/4-inch plywood on a slab. Use 1 to 1 1/4 inches in wood-joist construction or over screeds.

Some manufacturers recommend face nailing in addition to other fastenings. Another practice sometimes recommended is to leave a slight crack, about the thickness of a putty knife, between planks.

LAYING OVER OLD FLOORING

You can also lay a new strip floor over an old floor. In this case, the existing wood floor can serve as a subfloor.

Drive down any raised nails, renailing loose boards and replacing any warped boards that can't be made level. Sweep and clean the floor well, but don't use water. Remove thresholds to allow the new flooring to run flush through doorways. Remove doors and baseboard.

Lay asphalt felt or building paper over the old floor, as discussed earlier. Always install the new floor at right angles to the old floor boards. This method is sometimes the best way to handle an old hardwood floor in need of extensive repair.

PARQUET AND BLOCK FLOORING

The styles and types of block and parquet flooring, as well as the recommended procedures for application, vary somewhat among the different manufacturers. Detailed instructions on installation are usually provided with the flooring or are available from the manufacturer or distributor.

I'll cover the installation of parquet, block, herringbone, and similar flooring here and later in this chapter. This section applies only to tongue-and-groove parquet flooring where tongues and grooves are engaged. It doesn't apply to slat-type or finger-block parquet.

First, lay both blocks and the individual pieces of parquetry in mastic to a wood subfloor or over a moisture barrier, as described earlier. Use a cold, cutback asphalt mastic spread over the entire area to be floored at the rate of 1 gallon per 40 square feet. Use the notched edge of the trowel. Allow to harden a minimum of 2 hours or up to 48 hours, as directed by the manufacturer. The surface will be solid enough after 12 hours to allow you to snap working lines on it. Use blocks of the flooring as stepping stones to snap lines and begin the installation.

There are two ways to lay out parquet. The most common is with edges of parquet units (and thus the lines they form) square with the walls of the room. The other way is a diagonal pattern, with lines at a 45-degree angle to the walls.

Let's learn about the installation of the square pattern first. Never use the walls as a starting line because walls are almost never truly straight. Instead, use a chalk line to snap a starting line about 3 feet or so from the handiest entry door to the room, roughly parallel to the nearest wall. Place this

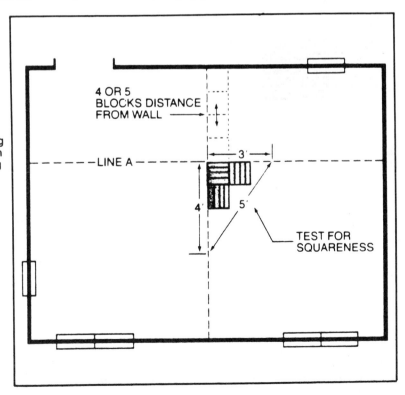

Fig. 3-24. Working lines for laying block flooring in a square pattern (courtesy National Oak Flooring Manufacturers Association).

4 OR 5 BLOCKS DISTANCE FROM WALL

LINE A

3'

4'

5'

TEST FOR SQUARENESS

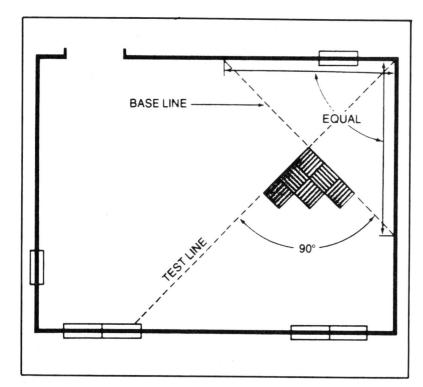

BASE LINE

EQUAL

TEST LINE

90°

Fig. 3-25. Working lines for laying block in a diagonal pattern (courtesy National Oak Flooring Manufacturers Association).

line exactly equal to four or five of the parquet units from the center of the entry doorway.

Next find the center point of this baseline and snap another line at an exact 90-degree angle to it from wall to wall. This will become your test line to help keep your pattern straight as the installation proceeds. A quick test for squareness is to measure 4 feet along one line from where they intersect, and 3 feet along the other. The distance between these two points will be 5 feet if the lines are true (Fig. 3-24).

Here's how to lay a diagonal pattern of parquet block flooring. Measure equal distance from one corner of a room along both walls, and snap a chalk line between these two points to form the baseline. This pattern need not be a precise 45-degree angle to walls in order to appear perfect. A test line should again intersect the center of the baseline at an exact 90-degree angle (Fig. 3-25).

Special patterns are also easy to install. Most exciting parquet patterns can be laid out with these two working lines just covered. Herringbone will require two test lines, however. One will be the

90-degree line already; the other crosses the same intersection of lines, but at a 45-degree angle to both (Fig. 3-25).

If such elaborate preliminary layout preparation seems a bit overdone, keep in mind that it is wood you are installing. Each piece must be carefully aligned with all its neighbors. Small variations in size, natural to wood, must be accommodated during installation to keep the overall pattern squared up. You cannot correct a "creeping" pattern after it develops. The more carefully laid-out floor causes less problems.

Wood parquet must always be installed in a pyramid or stair-step sequence, rather than in rows. This method again prevents the small inaccuracies of size in all wood from magnifying, or "creeping," until it looks misaligned.

Place the first parquet unit carefully at the intersection of the base and test lines. Lay the next units ahead and to the right of the first ones, along the lines. Then continue the stair-step sequence, watching carefully the corner alignment of new units with those already in place. Install in a quadrant of the room, leaving trimming at the walls until later. Then return to the base and test lines and lay another quadrant, repeating the stair-step sequence. Cover this area with the floor. Work backwards from the baseline toward the door. A reducer strip may be needed at the doorway.

Most wood floor mastics, regardless of type or open time, will allow the tiles to slip or skid when sidewise pressure is applied for some time after open time has elapsed. By working from *knee boards,* or plywood panels laid on top of the installed area of flooring, you avoid this sidewise pressure. For the same reason, no heavy furniture or activity should be allowed on the finished parquet floor for about 24 hours. Some mastics also require rolling.

Cut blocks or parquetry pieces to fit at walls, allowing 3/4-inch expansion space on all sides. Use cork blocking in 3-inch lengths between the flooring edge and the wall to permit the flooring to expand and contract (Fig. 3-26).

With blocks, a diagonal pattern is recommended in corridors and in rooms where the length is more than 1 1/2 times the width. This diagonal placement minimizes expansion under high humidity conditions.

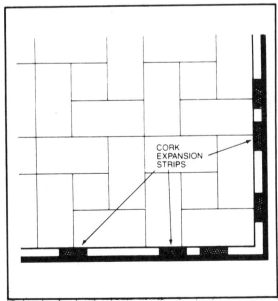

CORK
EXPANSION
STRIPS

Fig. 3-26. Use of cork blocking around edges of a block floor (courtesy National Oak Flooring Manufacturers Association).

SPECIAL INSTALLATION PROBLEMS

Every do-it-yourselfer would prefer "typical" installations with no handicaps or problems to overcome. Life doesn't seem to work this way however. Here are instructions on how to solve special hardwood flooring installation problems.

Oak Flooring over a Radiant-Heated Concrete Slab. Flooring will not impair the efficiency of the heating system, but slightly higher water temperatures may be required. An outside thermostat is therefore recommended to anticipate rapid temperature changes. Boiler water temperature must be controlled to keep it to a maximum of 125 degrees Fahrenheit and so limit the temperature of the slab surface to about 85 degrees, an acceptable level for most mastics.

The flooring is installed as in any other slab project, except do not fasten plywood to concrete with either nails or powder-actuated fasteners. Turn on the heating system for at least 48 hours before delivery of the flooring to the job because the heat will drive remaining moisture out of the slab. Allow flooring to become acclimated to the environment for 2 or 3 days, then install by the recommended slab methods covered earlier in this chapter. Remember to check flooring and mastic manufacturers' specifications for suitability for use over radiant heat.

Strip Flooring in a Wood Plenum System. This method of housing construction utilizes a crawl space that is completely sealed to the outside as a plenum to which air from the heating/cooling system is supplied. The air then enters each room through floor ducts.

A ground cover of polyethylene film is essential. The heating system must also operate for at least 48 hours prior to delivery of the flooring to stabilize the moisture condition. No other special consideration is necessary to install the flooring.

GYM FLOORING

Hardwood flooring is extremely popular for school, athletic clubs, and home gymnasiums. Gymnasium floor products are often made of 3/4-inch pecan or maple. Beach and oak are also suitable.

It is most important to have some resiliency built into these floors. In most respects, however, installation closely follows the screeds-in-mastic method recommended for conventional use, with a plywood or board subfloor installed over the screeds.

Make sure the slab is dry and level with a good float finish. Maximum surface variation is 1/4 inch in 10 feet. Grind down high areas and fill low areas with concrete leveling compound.

Sweep the slab clean and prime with asphalt primer. Let dry thoroughly and coat with asphalt mastic, using a notched trowel designed to apply at a rate of 40 square feet per gallon. Embed a layer of 15-pound asphalt felt or building paper, starting at a wall with a half sheet. Lap seams. Cover this with another layer of mastic and embed a second layer of asphalt felt or building paper, starting at the same wall with a full sheet to cover the seams of the first layer.

Either hot or cold mastic is satisfactory. If the cold type is used, be sure to allow 2 hours for solvents to evaporate before applying the building paper.

An alternate method of surface damp-proofing is to embed a 6-mil polyethylene film in a cold mastic, as described earlier. Lap the film edges 6 inches.

A suspended concrete slab needs no surface damp-proofing. Cross-ventilation below the slab is essential, however, and if over exposed earth, a ground covering of 6-mil polyethylene should be provided.

Screeds used and their application are identical to that previously described, with the following exceptions. Place them on 12-inch centers. If a subfloor is used, 16-inch centers are allowed. Leave a 2-inch space between the ends of the screeds and the base plate on all walls to allow for expansion.

The finish flooring may be nailed directly to the screeds. A much more sound and satisfactory floor can be achieved, however, by installing a subfloor of 3/4-inch minimum plywood or 3/4-inch dressed square-edge boards no wider than 6 inches. Follow the arrangement and nailing schedules described previously. If boards are used, leave 1/2-inch of space between them.

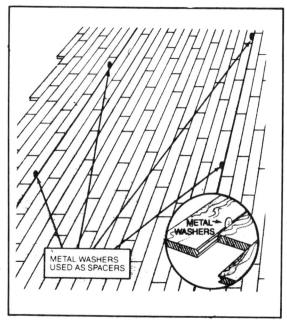

Fig. 3-27. Use of metal washers to provide expansion space on a gym floor (courtesy National Oak Flooring Manufacturers Association).

Start laying the finish flooring in the middle of the room and work toward the walls. Place the first two courses groove to groove with a slip-tongue joining the strips, and face-nail as well as blind-nail both courses. Proceed with succeeding courses in the conventional manner, using either 7d or 8d cut steel nails, screw-type nails, or 2-inch barbed fasteners.

After an area 3 to 4 feet wide has been laid across the room, leave a 1/16-inch expansion space between the last course laid and the next course (Fig. 3-27). Repeat the expansion space evenly at 3- to 4-foot intervals across the room.

Nailing is most important. Nail to all screeds and to both screeds when a strip passes over a lapped screed joint. All end joints do not need to meet over screeds, but adjoining strips should not break over the same screen space. If a subfloor is used, nails must be no more than 10 to 12 inches apart.

Allow a 2-inch expansion space along all walls and doorways. It can be covered at the walls with an angle iron bolted to the wall or a special wood molding, and at the doorways by a metal plate designed for such use.

MAKING INSTALLATION EASIER

The National Oak Flooring Manufacturers' Association recommends a number of tips for easier and better flooring installation.

Work from left to right. In laying strip flooring you'll find it easier to work from your left to your right. Left is determined by having your back to the wall where the starting course is laid. When it is necessary to cut a strip to fit to the right wall, use a strip long enough so that the cutoff piece is 8 inches or longer, and start the next course on the left wall with this piece.

Save short pieces for closets. For best appearances always use long flooring strips at entrances and doorways. Save some of the short pieces for closet areas and scatter the rest evenly in the general floor area.

Put a "frame" around obstruction. You can give a much more professional and finished look to a strip flooring installation if you "frame" hearths and other obstructions, using mitered joints at the corners.

Reverse direction of strip flooring. Sometimes it's necessary to reverse the direction of the flooring to extend it into a closet or hallway. To do this, join groove edge to groove edge, using a slip-tongue available from flooring distributors. Nail in the conventional manner.

Use only sound, straight boards for subfloors. The quality of the subflooring will affect the finish flooring. Use only square-edge, 3/4-inch dressed boards no wider than 6 inches. Boards that have been used for concrete form work are often warped and damp and should not be used.

Don't pour concrete after flooring is installed. Concrete basement floors are sometimes poured after hardwood flooring has been installed. Many gallons of water from drying concrete are evaporated into the house atmosphere, however, where it may be absorbed by hardwood flooring and other wood components. This is not a recommended building practice since excessive moisture will cause problems with wood floors and other woodwork. Wood flooring should not be installed until after all concrete and plaster work is completed and dry.

Put voids between screeds to good use. Masonry insulation fill, normally used in hollow concrete blocks, can be poured between the screeds or sleepers of a slab installation to give additional moisture protection and deaden the drumming sound that sometimes occurs from foot traffic.

Deaden sound in multi-story building. Noise transmission from an upper to a lower floor can be reduced in the following way. Nail subfloor to the joists in the normal manner and cover it with 1/2-inch or thicker cork or insulation board laid in mastic. Cover this with another 3/4-inch plywood subfloor, also laid in mastic. Nail the finish strip or plank floor to the plywood, or lay block or parquetry floors in mastic on the plywood. In the case of parquet, the second subfloor plywood can be a 1/2-inch tongue-and-groove type. Note that specifications for some high-rise apartment buildings call for other types of sound-deadening construction and should be followed.

Mastics and trowels. There are several types of mastic available that are satisfactory for use in laying hardwood floors. Hot asphalt is generally used only for laying screeds on concrete, and the screeds must be positioned immediately on pouring the mastic. Cutback asphalt, chlorinated solvent, and petroleum-based solvent mastics are all applied cold and are always used for laying block and parquet floors. Cutback asphalt is also used to hold a membrane damp-proofing. Follow the manufacturer's instructions on coverage, drying time, and ventilation.

Trowels usually have both straight and notched edges. The notched edge is for use where a correct mastic thickness is specified. Both mastic and trowels are available from flooring manufacturers and distributors.

FLOATING FLOOR INSTALLATION

As mentioned earlier, there are numerous ways of installing a satisfactory hardwood floor. Thus far in this chapter you've learned about the more traditional methods. Another method is called the *floating floor* system.

Figures 3-28 and 3-29 illustrate a cross section of a floating floor installation over wood or

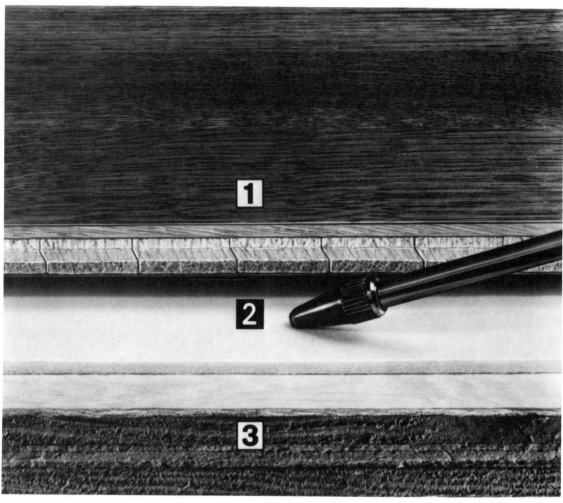

Fig. 3-28. Cross section of floating floor installation of (1) plank, (2) foam underlayment, and (3) plywood subfloor (courtesy Harris-Tarkett, Inc.).

Fig. 3-29. Floating floor over wood or suspended concrete (courtesy Harris-Tarkett, Inc.).

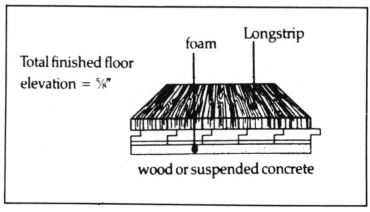

Total finished floor elevation = ⅝"

foam

Longstrip

wood or suspended concrete

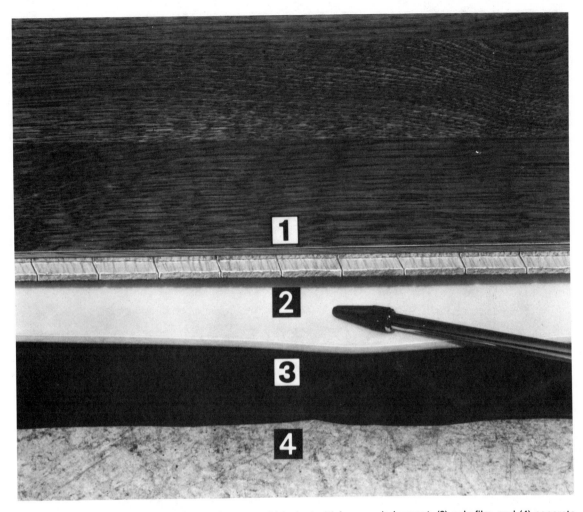

Fig. 3-30. Cross section of floating floor installation of (1) plank, (2) foam underlayment, (3) poly film, and (4) concrete (courtesy Harris-Tarkett, Inc.).

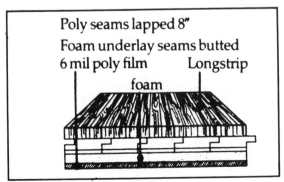

Fig. 3-31. Floating floor over concrete on-grade (courtesy Harris-Tarkett, Inc.).

suspended concrete using a foam underlayment. If it is in contact with the ground (on-grade), a 6-mil polyethylene film barrier should first be installed over the subfloor and the foam applied over the poly film (Figs. 3-30 and 3-31). Lap seams of poly film at least 8 inches over each other. If the subfloor is not level to within 3/16 inch in a 10-foot radius, it should be leveled with a latex leveling compound.

Tools required for a floating installation include a hammer, a hand or power saw, wood or plastic wedges (or equivalent 1/2-inch spacers), a crowbar, a chalk line, installation adhesive, a tapping cover block, and any special tools.

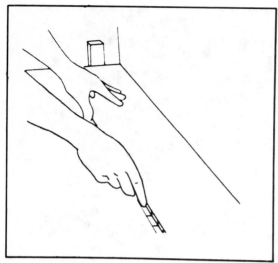

Fig. 3-32. Start at one sidewall (courtesy Harris-Tarkett, Inc.).

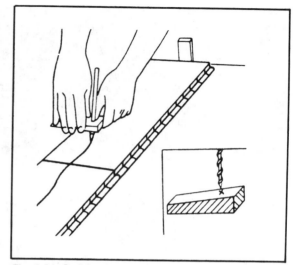

Fig. 3-34. Scribing boards (courtesy Harris-Tarkett, Inc.).

Once the poly film (if used) and the foam underlayment have been installed over the subfloor, the job site is ready for the boards. Don't open the bundles until you are ready to begin the installation process. Decide which direction the boards will run. Start at one side wall (Fig. 3-32) with the first row of boards. Allow a 1/2-inch expansion along side and end walls (Fig. 3-33) by using wood or plastic wedges (equivalent spacers). If the starting wall is out of square, it is recommended that the first row of boards be scribed to allow for 1/2-inch of expansion (Fig. 3-34) and a straight working line.

The specially milled boards must be edge- and end-glued using the manufacturer's adhesive or equivalent. Apply in 8-inch long beads with 12-inch spacing between beads (Fig. 3-35) along the side

Fig. 3-33. Use wedges to establish expansion space (courtesy Harris-Tarkett, Inc.).

Fig. 3-35. Applying adhesive along side grooves (courtesy Harris-Tarkett, Inc.).

Fig. 3-36. Applying adhesive on end joint (courtesy Harris-Tarkett, Inc.).

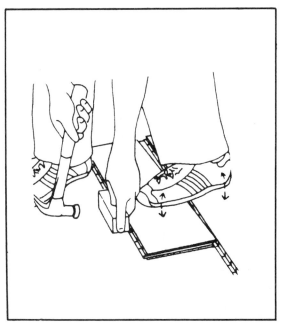

Fig. 3-38. Installing subsequent rows with a hammer and block (courtesy Harris-Tarkett, Inc.).

grooves. Fully glue every end joint (Fig. 3-36). If any excess glue squeezes up to the finish surface, wipe off with a moist cloth.

Install the first row, using the appropriate expansion space with the groove side facing the wall (Fig. 3-37). The subsequent rows are installed, edge- and end-glued, with a hammer (Fig. 3-38) and tap-

ping block to prevent damage to the protruding tongue. Check for tight fit on sides and ends and stagger 2 feet between the end joints of adjacent board rows (Fig. 3-39).

Most often, the last row does not fit in width. When this occurs, follow this simple procedure. Lay a row of boards, unglued, and tongue toward the

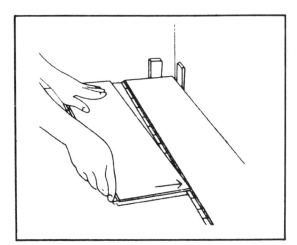

Fig. 3-37. Installing first row (courtesy Harris-Tarkett, Inc.).

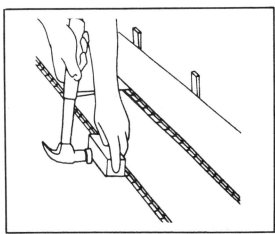

Fig. 3-39. Check for tight fit of boards (courtesy Harris-Tarkett, Inc.).

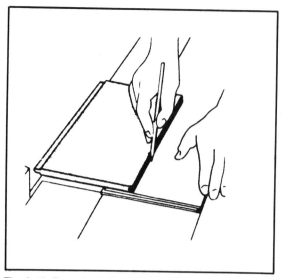

Fig. 3-40. To make last row fit in width: first, lay a row of boards on top of the last installed row (courtesy Harris-Tarkett, Inc.).

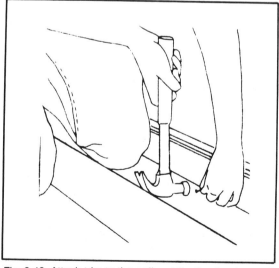

Fig. 3-42. Wedge the last row of boards into place with a crow bar (courtesy Harris-Tarkett, Inc.).

wall, directly on top of the last installed row (Fig. 3-40). Take a short piece of the strip flooring with the face down and tongue side against the wall. Draw a line with a pencil along the row moving down the wall. The resulting line gives the proper width for the last row which, when cut (Fig. 3-41), can then be wedged into place using a crowbar and cover board (Fig. 3-42) to prevent damage to finished walls or molding.

Make sure that wedges or spacers are removed when the installation is complete. The expansion space should also be covered with an appropriate molding that allows the boards to move freely underneath. Always attach the trim to the wall or

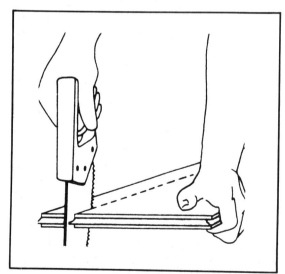

Fig. 3-41. Cut the board on the scribed line (courtesy Harris-Tarkett, Inc.).

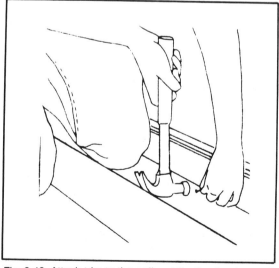

Fig. 3-43. Attach trim to the wall, not the flooring (courtesy Harris-Tarkett, Inc.).

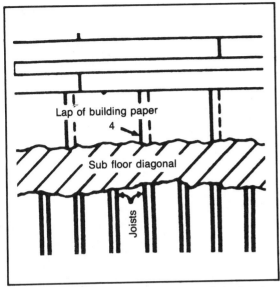

Fig. 3-44. Cross section of plank flooring installation (courtesy Harris-Tarkett, Inc.).

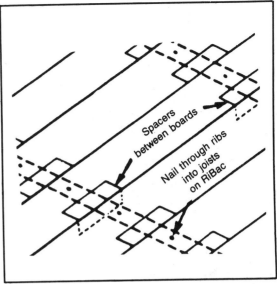

Fig. 3-46. Using spacers between plank flooring (courtesy Harris-Tarkett, Inc.).

vertical object (Fig. 3-43) and never to the floor boards.

One more note: in large areas measuring more than 24 linear feet, use 1/4-inch expansion for each 12-linear-foot width and length.

INSTALLING PLANK FLOORING

General installation of plank flooring was

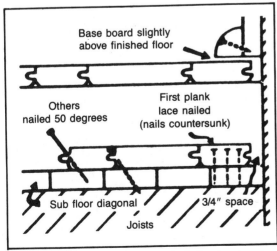

Fig. 3-45. Installation details of plank flooring (courtesy Harris-Tarkett, Inc.).

covered earlier in this chapter. Here's another popular method.

Suggested tools and accessories include a conventional claw hammer or power nailer and a No. 80, 3/4-inch counterbore, unless the flooring is prebored at the factory. You'll also need 7d, 2 1/4-inch, screw-type flooring nails for conventional nailing or power cleats for power nailing. No. 9, 1 1/4-inch flathead wood screws and walnut or oak wooden plugs, 5/16-×-3/4-inch diameter, should be used.

Figure 3-44 illustrates a cross section of plank flooring installation, including the subflooring. Kiln-dried coniferous lumber, 1-×-4-inch to 6-inch wide, square-edged, and laid diagonally over 16-inch on-center wooden joists is recommended. A minimum 1/2-inch thick exterior plywood can also be used with long edges at right angles to joists and staggered so that the end joints on adjacent panels break over different joists. Particleboard is not considered a suitable subfloor.

Refer to Figs. 3-45 through 3-48. Lay out your plank flooring only after the plasterboard and tile work have thoroughly dried and all but the final woodwork and trim have been completed. The building interior should have been dried and

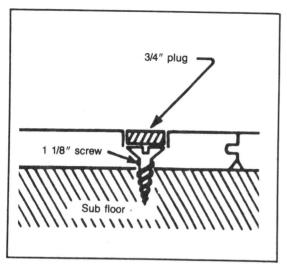

Fig. 3-47. Installation of wood screw and plug to fasten plank flooring (courtesy Harris-Tarkett, Inc.).

into the joists. The number of screws to use is a function of job conditions and the desired surface appearance. Use one screw and plug on each end of 3- and 4-inch wide planks, two widthwise on the ends of 5- and 6-inch planks, and three widthwise on the ends of 7- and 8-inch planks.

Sanding and finishing your plank flooring will be covered in Chapter 4.

INSTALLING PARQUET FLOORING

One of the easiest and most beautiful flooring you can install is the parquet floor (Fig. 3-49). Early parquet floors were extremely difficult to install correctly, but modern manufacturing and application techniques have placed parquet flooring within the "can-do" realm of the typical do-it-yourselfer.

Preparation of the subfloor for the installation of parquet flooring is the same as for other types of hardwood flooring. Make sure it's nailed down tightly, clean, and level. If you've stripped an old floor off to install your parquet flooring, make sure that all traces of the old flooring, mastic, and fasteners are gone. Old wood should be thoroughly sanded with 3 1/2 open grit paper to remove oils, waxes, paints, varnishes, glues, and other foreign

seasoned, and a comfortable working temperature should exist during the plank flooring installation.

Make sure that the subfloor is adequately and properly nailed before you start. Clean the subfloor surface and cover it with 15-pound or 30-pound asphalt saturated felt. Lap the edges at least 4 inches. Double the felt around heat ducts in the floor.

Plank flooring should be laid at right angles to the floor joists and, if possible, in the direction of the longest dimension of the room. Begin laying tongue-and-groove plank flooring in a room corner with the edge groove of the planks facing the wall. Provide no less than 3/4-inch expansion space or what will be covered by the baseboard and trim.

The first run of planks should be face-nailed, then countersunk. All other runs should be nailed at a 50-degree angle on 8-inch centers at the tongue. Make sure the end joints are staggered. Remember to leave a small space between the planks—usually 1/32 inch is sufficient.

After nail installation of the plank flooring, bore appropriate holes for screws and plugs. Use the counterbore if the flooring was not factory bored or if additional boring is desired. Screws and plugs should be located at the ends of each plank and at intervals frequent enough to hold the planks securely. Whenever possible, screws should extend

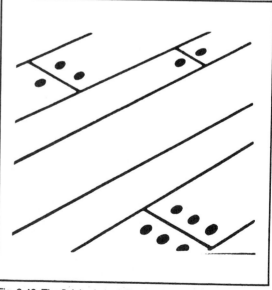

Fig. 3-48. The finished plank flooring (courtesy Harris-Tarkett, Inc.).

Fig. 3-49. Parquet block flooring is actually easy to install (courtesy Pennwood Products Co.).

matter. After sanding, the subfloor should be swept and vacuumed to remove dirt. An alternate method is to cover existing subfloor with 1/2-inch plywood.

Figure 3-50 shows the proper method of laying out the room prior to installation and the proper procedure to follow in laying the parquet blocks. Locate the center of the room from the length side and width side with a chalk line. Then determine the number of parquet blocks to be laid from the center of the room to the wall. If the measurement leaves less than 2 inches of a block to the wall, shift the centerline to allow 2 inches or more of the block to be laid at the wall. Be sure that at least 1/2 inch or more of expansion space is left around all four walls. Following installation, cover the expansion area with baseboard or quarter-round.

Spread the adhesive according to label instructions. Lay the block onto a quadrant of the room from the centerline, completing that area before starting the next. Accurate alignment of the first 10 blocks is most important. Never force or tap blocks

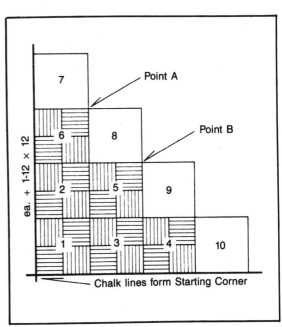

Fig. 3-50. Laying out a parquet block floor (courtesy Pennwood Products Co.).

with a hammer. Just lift the tongue-and-groove blocks tightly by hand. Be sure that corners form right angles and match properly at points A and B in Fig. 3-50.

ADHESIVE

Adhesive is used in most hardwood flooring installation. It is especially critical to the installation of parquet block flooring.

Spread the adhesive on the floor with a notched trowel. Apply the adhesive to an area of such size that flooring can be laid into adhesive within approximately 2 hours. Don't let a heavy skin form. A light skin on the adhesive can easily be broken by pressing the flooring into it.

After spreading the adhesive, wait to install the flooring—leave the adhesive "open"—to help ensure a faster bond. Allow 20 to 30 minutes, depending on the temperature and humidity.

To install the flooring, press down firmly. Lift a piece of flooring occasionally to be sure the adhesive is transferring to the back. Remember that wood flooring needs expansion space around its edges.

Once installation is complete, some adhesives require that you roll the floor carefully in both directions using a heavy roller of 75 to 100 pounds. To remove adhesive from tools, use naptha, white gas, or mineral spirits, being careful not to get it on surface it could damage.

LEAVING YOUR MARK

Since hardwood floors are designed and installed to last a long time, some installers will date and "sign" a final strip, plank, or block before installing it (Fig. 3-51). It can then be placed in a corner, under a molding, or on a door threshold. Such a stamping kit can be rented, borrowed, or purchased from a variety of sources, including many

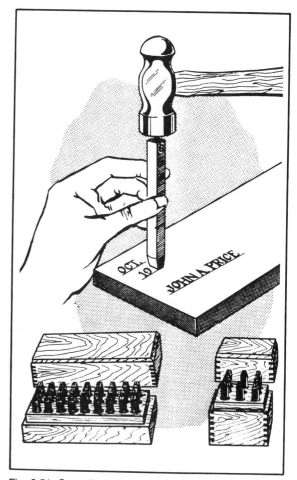

Fig. 3-51. Stamping a corner board.

woodworking shops or even the supplier of your hardwood flooring.

Your hardwood floor is installed! Congratulations. The work isn't over, though. The final step in the installation of a hardwood floor is the finishing, which includes sanding and then applying one of the many types of finishes that will ring out the wood's natural beauty while protecting it from wear.

Chapter 4

Finishing Hardwood Floors

WHILE SOME HARDWOOD FLOORING CAN BE purchased prefinished, the majority must be sanded and finished once installation is completed. In this chapter you will learn how to finish your new hardwood floor using methods and equipment developed by professional hardwood flooring installers over many years. Chapter 6 will cover the refinishing of old hardwood floors.

Finishing your hardwood floor requires some specialized and expensive power tools. All can be rented, however, at most rental yards or from the larger flooring supply houses. Some suppliers will even loan you equipment free or at a nominal charge if you purchase the majority of your flooring needs through them. These primary pieces of floor finishing equipment include the drum sander, the power edger, and the floor polisher. Finish application equipment can also be rented or borrowed.

FLOOR PREPARATION

Applying the finish to the hardwood floor should be one of the last jobs of any construction project. In this way other work and the traffic of workers won't mar the finish. Wall coverings should be in place and painting completed except for a final coat on the base molding. Sweep the floor clean immediately before sanding.

SANDING

A drum-type floor sander (Fig. 4-1) is used for heavy sanding operations. A floor polisher with a sanding or screen disc or steel wool is used for special situations and to give the floor an extremely fine finish. Professional floor finishers will also use a spinner-type edger in areas the drum sander can't reach. Since the spinner-type edger is difficult to use, however, you may elect to hand-scrape and hand-sand around the edges of the room.

Load the drum sander with a medium abrasive (Fig. 4-2). Place the machine along the right-hand wall with about 2/3 the length of the floor in front of you. Start the motor and ease the drum to the floor. Walk slowly forward, letting the machine pull you along at an even pace. As you near the wall,

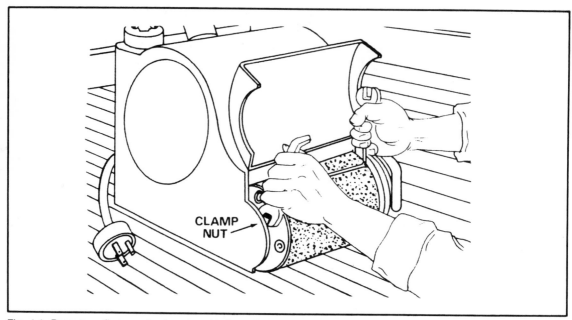

Fig. 4-1. Drum-type floor sander (courtesy National Oak Flooring Manufacturers Association).

gradually raise the drum off the floor by lifting up on the control handle.

Start pulling the machine backward and ease the drum to the floor. Cover the same path as you made on the forward cut. When you reach your starting point, ease the drum from the floor and move the machine to one side approximately 4 inches. Then repeat the forward and backward passes.

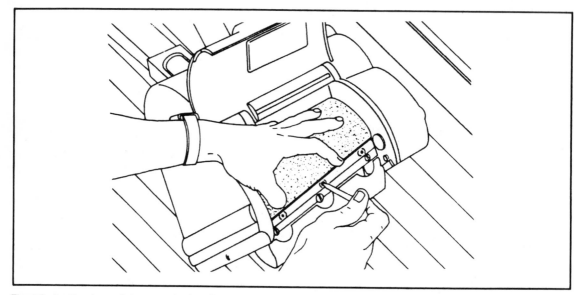

Fig. 4-2. Another type of drum sander has the paper held in place by a clamp (courtesy National Oak Flooring Manufacturers Association).

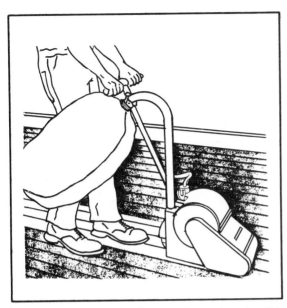

Fig. 4-3. Always keep the drum sander moving when it is operating (courtesy National Oak Flooring Manufacturers Association).

When 2/3 of the room is sanded, turn the machine in the opposite direction and sand the remaining third in the same manner. Be sure the cuts made in the 1/3 area overlap the first cuts by 2 to 3 feet. This method blends the two areas together.

It is very important that you never let the sanding drum touch the floor unless you are moving the machine forward or backward (Fig. 4-3). If you do, it will cut a hollow in the floor that cannot be removed.

After you finish the first cut with the drum sander (Fig. 4-4), use the power edger (Fig. 4-5) or scraper and hand-sand up to the baseboard, in corners, closets, and other areas the drum sander won't reach. Use the same grit as used on the drum sander, moving the edger in a brisk, left-to-right, semicircular pattern.

When using a hand scraper (Fig. 4-6) instead of the power edger, apply even pressure, scraping in the direction of the grain or from the wall into the sanded area if the floor is parquet. Avoid gouging the wood with the scraper. After scraping use a sanding block and paper with the same grit as used on the drum sander to smooth the flooring. A brick with a piece of old blanket glued around it makes a good sanding block.

After you have sanded the entire floor with medium abrasive, repeat the entire procedure with fine abrasive (Table 4-1). Sand the body of the floor first and then the edges, paying particular attention to blending the edges with the main floor area.

Often only two sanding cuts are used, but for

Fig. 4-4. Finishing the first cut with the drum sander (courtesy National Oak Flooring Manufacturers Association).

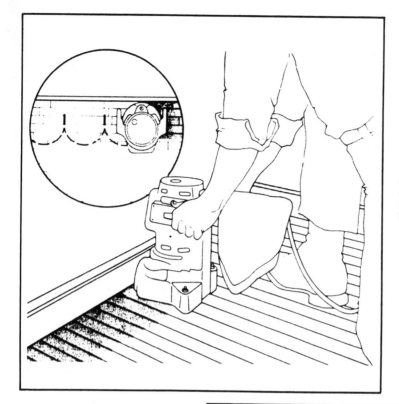

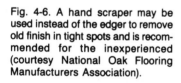

Fig. 4-5. Use the power edger to sand up to the baseboard and in other areas the drum sander won't reach (courtesy National Oak Flooring Manufacturers Association).

Fig. 4-6. A hand scraper may be used instead of the edger to remove old finish in tight spots and is recommended for the inexperienced (courtesy National Oak Flooring Manufacturers Association).

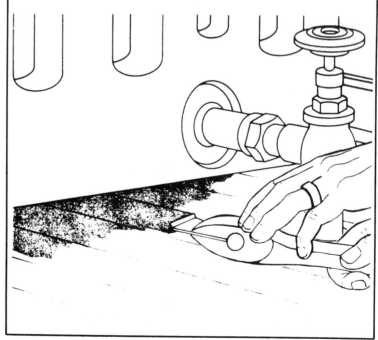

FLOOR	OPERATION		GRADE OF SANDPAPER	
Hardwood Oak, Maple, Beech, Birch	First Cut	Uneven Floor	Medium-Coarse	2 (36)
		Ordinary Floor	Fine	1 (50)
	Final Sanding		Extra Fine	2/0 (100)
Softwood Pine Fir	First Cut	Uneven Floor	Medium-Fine	1½ (40)
		Ordinary Floor	Fine	1 (50)
	Final Sanding		Extra Fine	2/0 (100)

Table 4-1. Sanding New Floors.

a smoother, finer finish, switch from the drum sander to the floor polisher fitted with fine paper or a screen disc and sand the entire floor. If the floor is to be stained, however, use a slightly heavier grit on this cut to leave a "tooth" on the wood surface which enables stain to penetrate more readily.

Sanding Strip and Plank Flooring

If the floor is flat and level, make all cuts parallel to the direction of the strips (Fig. 4-7). If the floor is uneven, however, make the first cut at a 45-degree angle to the direction of the strips. This will remove any peaks or valleys caused by minute variation in thickness of the strips or in the subfloor. Make succeeding cuts parallel to the direction of the strips. Always use at least two cuts. A third and even a fourth cut with the floor polishing machine and sanding disc is recommended because of the fine finish it imparts to the floor.

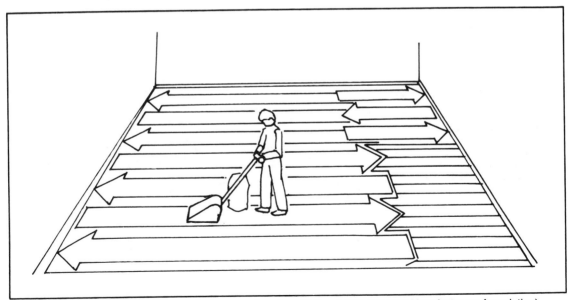

Fig. 4-7. Sand in the longest direction of the room (courtesy National Oak Flooring Manufacturers Association).

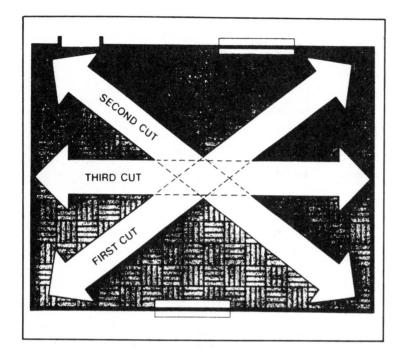

Fig. 4-8. Parquet, block, herringbone, and similar flooring should have three sanding cuts, as shown (courtesy National Oak Flooring Manufacturers Association).

Sanding Parquet, Block, and Similar Flooring

Use the drum sander for the first two cuts. Make the first cut with medium-grit paper on a diagonal of the room, and the second cut with fine-grit paper on the opposite diagonal. Then switch to the floor polishing machine and, using a screen disc, make a third cut lengthwise of the longest room dimension (Fig. 4-8).

PREPARING FOR THE FINISH

When sanding is completed, sweep or vacuum the floor clean. Wipe up all dust on windows, sills, doors, door frames, baseboards, and the floor using a painter's tack rag.

Inspect the floor carefully and hand-sand to remove any scratches or swirl marks. Fill cracks and nail holes with a commercial putty of a matching color which is compatible with stain and/or finish, or make your own putty with dust from the fine sanding mixed with a floor sealer to form a thick paste. Apply with a putty knife and scrape off the excess. When dry, hand-sand with fine paper. When these operations are finished, both old and new floors become essentially new wood surfaces, and should be treated as new floors.

Apply the first coat of stain or other finish the same day that sanding is completed to help keep the wood grain from rising and creating a rough surface. With stain and all finishing materials, be sure to read and follow the label instructions. Avoid vig-

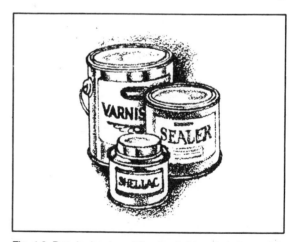

Fig. 4-9. Popular hardwood flooring finishes include varnish, sealer, and shellac (courtesy National Oak Flooring Manufacturers Association).

Fig. 4-10. Hardwood flooring finishes are often applied using a paint brush (courtesy National Oak Flooring Manufacturers Association).

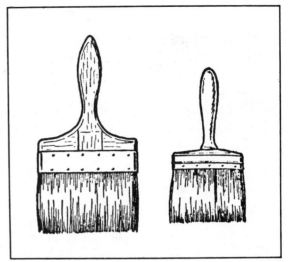

Fig. 4-12. Two large flat brushes.

orous shaking or stirring, which may cause air bubbles to be trapped in the material and affect the quality of the finish. Allow adequate ventilation and avoid prolonged breathing of fumes.

Finishing your hardwood floor involves numerous products, like sealer, stains, shellac, and varnish (Fig. 4-9) and applicating equipment (Figs. 4-10 through 4-14). To guide you, Table 4-2 suggests numerous types of finishes for specific types of woods used in home construction.

COLORING THE FLOOR

If other than a natural finish is desired, coloring must be the first step in the finishing process. Let's first look at coloring sealers and stains.

Colored Penetrating Sealers. These are generally available in a number of wood tones and several other colors. They offer an easy method since the floor is colored as part of the finishing operation. Their application will be covered later.

Pigmented Wiping Stains. These are penetrating oil-resin vehicles to which pigments have been added. The pigments are not in solution, but in suspension; so the material must be stirred regularly during use to maintain a uniform color. The pigment collects in the open pores of the wood

Fig. 4-11. Typical varnish brush.

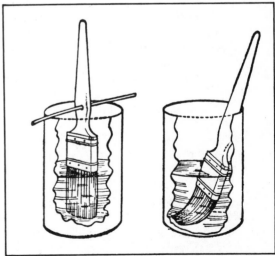

Fig. 4-13. Right and wrong way to clean brushes in solvent.

Table 4-2. Home Interior Woods.

Type	Characteristics
American Woods Hard: Birch	Takes all kinds of stains well; no filler required; used for veneers, to imitate other woods, for trim, interior finishings, furniture; good for blonde finishes
Cherry	Takes light reddish stain; generally no filler needed unless lacquered; will not bleach; used for furniture, to imitate mahogany
Hickory	Takes water stain best; takes fine polish after sanding; generally a heavy filler needed; used for furniture, to imitate mahogany and walnut
Maple	Stains well in lighter colors; takes fine finish; no filler needed; widely used for furniture, veneers, floors
Oak	Takes many kinds of stains; heavy filler generally needed, but not for many period effects; widely used, especially for floors and heavier pieces of furniture
Walnut	Takes many stains well; bleaches well; heavy filler generally needed, except for old English effect; used for furniture, veneers
Soft: Cypress	Takes oil and water stains well; close, even grain; used for interior and exterior trim, garden furniture
Fir	Takes oil stains well; has very strong figure; used for interior trim; doors, flooring, also shelves, cupboards
Pine, yellow	Takes oil stain well; prominent grain; resinous; widely used for bedroom and kitchen furniture, closets, bookcases, floors
Pine, white	Takes oil and water stains well; no filler needed generally; straight grained; when finished is generally used to imitate maple
Redwood	Takes red stain well; does not readily bleach or take lighter stains; filler sometimes used; used for veneers
Imported Woods Avodire	Natural finish preferable; takes light stain; requires filler; bleaches well
Circassian Walnut	Stains well; bleaches well; requires filler; has very fancy figure; used mostly for veneers
Mahogany, Philippine	Takes all stains well; needs heavy filler; has very beautiful grain; used for furniture, trim, floors, to imitate walnut and true mahogany; may fade if not properly finished
Mahogany, true	Takes all stains well; generally needs filler; excellent color and figure; used for furniture; often called the finest of all woods
Rosewood	Requires "washing" before finishing; needs filler; will not bleach; used mostly for veneers
Satinwood	Takes light stains; often finished naturally; needs filler; will not bleach; used mostly for veneer and inlay

and thus accentuates the grain pattern.

Apply by brushing on a generous coat. Allow the pigment to penetrate and then wipe it off with clean rags. Wipe vigorously to remove as much stain as possible from the surface; the pigment remaining in the pores and grain lines provides the color. Always test first because the length of time the stain is allowed to remain on the floor will determine the degree of color tone provided.

Varnish Stains. These are a formulation of varnish with an oil-soluble dye to provide color. As with penetrating seals, you color the wood at the same time you apply the finish coat. Their main disadvantage is that they tend to obscure the wood grain to a great extent and thus do not provide nearly as beautiful a finish as one of the other methods.

TYPES OF FINISHES

Let's consider the properties and the method of application for the more popular finishes used with hardwood floors: penetrating seal, surface finishes, polyurethane, varnish, shellac, lacquer, and bleaching.

Penetrating Seal

This is the finish recommended for most residential floors. The sealer soaks into the wood pores and hardens to seal the floor against dirt and most stains. It wears only as the wood wears and will not chip or scratch. After years of wear the floor can usually be refinished without sanding by cleaning it and applying another coat of sealer or a special reconditioning product. Limited areas of wear can be refinished without showing lap marks where the new finish is applied over the old.

Penetrating seal can also be used as an undercoat for varnish or shellac when a high gloss is desired. It is available in natural and a number of wood tones and other colors.

There are two basic types of sealers, distinguished by drying time requirements. Normal or slow-drying sealers can be used safely by anyone. Fast-drying sealers should be used only be a pro-

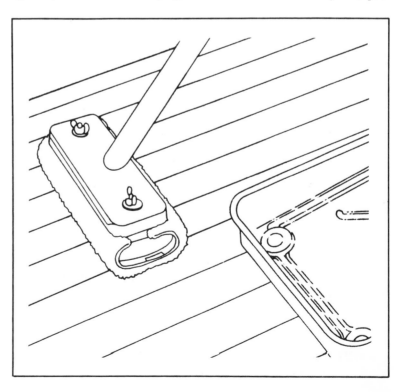

Fig. 4-14. A long-handled lamb's wool applicator is excellent for applying penetrating seal finishes (courtesy National Oak Flooring Manufacturers Association).

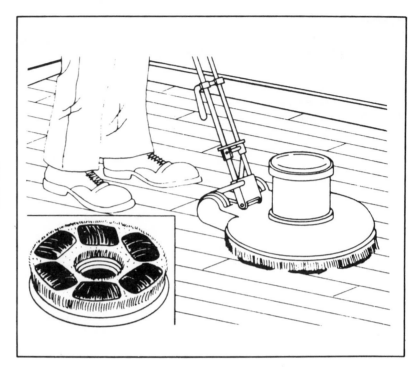

Fig. 4-15. Ordinary steel wool pads may be used with the floor polisher when buffing a sealer finish (courtesy National Oak Flooring Manufacturers Association).

fessional who is accustomed to handling and applying them and can complete the job within the allotted drying time to avoid lap marks or a splotchy appearance. Some seals produce satisfactory results with one coat, but most manufacturers recommend two coats or one plus a special top dressing.

Sealers are mopped on. Use a clean string mop or long-handled applicator with a lamb's wool pad (Fig. 4-14). Second choices are a wide brush or squeegee. Apply generously and wipe up excess with clean cloths or a squeegee. Note: some manufacturers call for rubbing the sealer into the wood with steel wool pads on an electric buffer while the sealer is still wet.

Allow to dry, then buff with No. 2 steel wool using an electric polisher equipped with a steel wool pad (Fig. 4-15) or buff by hand. For a stain-finish floor buff both coats with steel wool, then apply wax to protect and add more luster to the finish.

For even greater gloss and maximum protection from wear, use a single application of a penetrating seal and buff with steel wool. Follow this with two coats of polyurethane, to be covered later.

Surface Finishes

All of these finishes are applied with a high-quality brush or lamb's wool applicator designated for use with the particular type of finish selected.

On strip and plank flooring, work in the direction of the boards in a path narrow enough to keep a wet working edge as you move across the room (Fig. 4-16). With block and other pattern floors, apply the finish in strips the width of one block, or about 12 inches for other types, working from wall to wall across the shorter room dimension.

Lap strokes by working from the wettest area back into that just covered. Follow the manufacturer's directions for drying time between coats.

Polyurethane

These blends of synthetic resins, plasticizers, and other film-forming agents produce an extremely durable surface that is moisture-resistant. They are the best choice for kitchens, commercial applications, and other areas exposed to stains, spills, and high traffic.

There are two types: oil-modified polyurethanes are air-drying materials; solvents pass off in vapors.

The second type is the only one recommended for use by the nonprofessional.

Moisture-cured polyurethanes absorb minute quantities of moisture from the air, which causes them to cure and harden. While slightly more wear-resistant than the air-drying type, the moisture-cure polyurethanes are extremely difficult to apply properly. Both relative humidity (moisture in the air) and drying time are critical.

Somewhat similar to polyurethanes are the epoxy and urea-formaldehyde finishes, which are also synthetic products with high durability. Type of undercoat, working time, number of coats, and other factors are all critical and make application difficult. They should only be applied by the highly skilled.

Caution: adhesion of polyurethanes, epoxies, and urea-formaldehyde finishes is affected by wax and grease, as well as some types of stains, bleaches, or sealers. Further more, one type of polyurethane may not be compatible with another type. Always make a test in a closet or some other inconspicuous place to be sure the finish will adhere and dry properly. This is particularly important when you are refinishing an old floor since some of the old finish may have penetrated the wood fibers below the level to which it is sanded.

Use a brush or lamb's wool applicator to apply, working along the grain to get a smooth, even coat. Allow drying time as specified by the manufacturer, then buff with steel wool. Dust thoroughly and apply a second coat along the grain of the wood. The final coat does not require buffing.

For a particularly good finish combining the best qualities of a sealer and a surface coating, use a penetrating sealer followed by one or more coats of polyurethane. Check product labels to be sure the sealer and top coat are compatible.

Varnish

Use only varnish manufactured for flooring applications since it is more durable than ordinary varnish. Varnish gives a glossy finish of good durability and resistance to stain and spots. It shows scratches, however, and worn spots are difficult to patch without showing lines between the old and new finishes.

Varnish drying time is 8 hours or more. Use three coats over bare wood, or two coats over a wood filler or a shellac or sealer undercoat.

Thin clear varnish with one part thinner to eight parts of varnish for the first coat over bare wood. Use full strength over penetrating stain, shellac, or colored varnish. Apply all succeeding coats full strength, observing drying times recommended by the manufacturer. Remember that varnish takes much longer to dry than shellac or lacquer.

Flow varnish on as smoothly as possible, brushing out each stroke. Sand with fine paper between coats. Dust well and wipe with a turpentine-dampened rag between coats. Use two or three coats.

Shellac

Shellac is easy to apply and dries fast. Two or three coats can be applied in one day. It has moderate resistance to water and other staining agents, but will spot if liquids are not wiped up promptly. It gives a high gloss and will not darken with age as quickly as varnish. Use at least two coats and preferably three. Shellac may be used as a sealer under a varnish top coat.

Its main disadvantage is that it chips easily. Repairs can be made by sanding the chipped or worn area and touching up with new shellac. Be careful to "feather" the new finish into the surrounding area.

Use a 3-pound cut (3 pounds of shellac dissolved in 1 gallon of denatured alcohol). Apply with full, smooth flow using a wide brush and avoid puddling. Allow to dry 2 hours and then sand with fine paper. Dust and recoat, allowing 3 hours drying time before applying a third coat. Sand with fine paper then dust before the third coat. Use extra-fine paper to smooth the final coat. Allow to harden overnight before using.

Lacquer

Lacquer provides a glossy finish with about the same durability as varnish. An advantage is that worn spots may be retouched since the new lacquer dissolves the old and blends with it rather than forming an additional layer. Lacquer is difficult to apply because it dries so quickly, causing lap marks. Use by other than skilled applicators is not recommended. Two coats are required.

Apply lacquer with a wide brush or mohair roller. Work fast to prevent lap marks. Allow the first coat to dry 1 hour, hand-sand with fine paper, and dust. Apply the second coat and allow overnight drying before use. Do not sand the final coat. For a three-coat job, allow an hour for the second coat to dry, sand with fine paper, dust, and apply a third coat reduced with one part of thinner to four parts lacquer. Be sure to use a thinner recommended by the lacquer manufacturer to avoid drying problems.

Bleaching

An interesting effect can be obtained on both new and old floors by bleaching the wood to remove the color without obscuring the grain pattern. The general tone of the wood is retained, but the color intensity is reduced.

Wood bleaches are available at most paint stores. The best are usually two-solution products. The first solution is brushed on and changes the pigments chemically, while the second solution removes the pigments. Before using a bleach, be sure that the flooring is clean and free from oils, grease, old finish, and any dirt that might repel the bleach and prevent it from working or give it an uneven effect.

Since bleaches have a water base, their use will cause the grain of the flooring to raise. Sanding with fine paper after use will be necessary to restore the floor to the smooth surface required.

Follow manufacturer's instructions for application times on the bleach you use. It's a good idea to make a test using a piece of scrap flooring because the length of time the bleach remains on the floor will affect the degree of color removal.

A polyurethane finish may be applied to the bleached floor. Some finishers prefer, however, to follow the bleaching operation with a white stain or a wood filler mixed with wood pigment, which gives the whitest floor of all. Then use polyurethane for the final finish. Check the product label to be sure

Fig. 4-16. Work in the direction of the boards (courtesy National Oak Flooring Manufacturers Association).

it is compatible with the stain or pigmented filler used and will not turn an amber color over the whitened floor.

STENCILING

Decorative borders or overall floor patterns can be applied to hardwood floors with the use of stencils and masking tape available at many paint and decorating stores. The stencils or tape must be securely adhered to the finished floor surface to prevent the final color from bleeding beyond the area to be treated. Refer to Fig. 4-17.

One method is to use a clear or light-colored penetrating sealer over the entire floor area and a darker color sealer to create the pattern. For a more vivid color and pattern, use a good-quality floor enamel. Always make a test panel first to be sure of getting the desired effect and, if using paint, that it will adhere satisfactorily to the base finish.

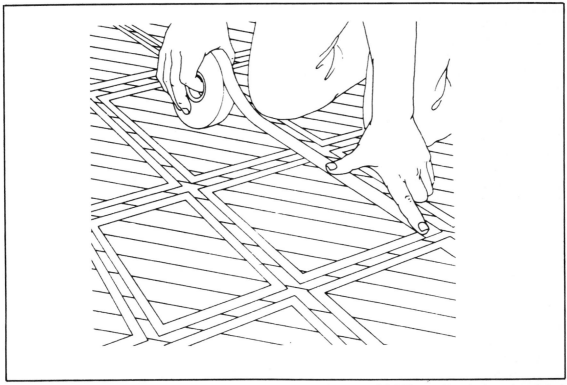

Fig. 4-17. Many interesting effects may be obtained by stenciling (courtesy National Oak Flooring Manufacturers Association).

Use paint sparingly and in other than high-traffic areas because it will chip and scratch easily. It should always be protected with one of the surface finishes described earlier to help prevent such damage.

USE OF FILLERS

Most hardwoods have minute crevices that are exposed by sawing and sanding. Before the development of modern finishing products, it was common to use a wood filler to fill these crevices or pores. Penetrating seals and polyurethane finishes do not require their use, and fillers are quite often omitted with other finishing materials.

There are, however, two situations when use of a filler is helpful. One is when a mirror-smooth finish is desired, usually with varnish as the top coat. By filling the pores of the wood, you get an absolutely smooth surface that, with a gloss varnish, has extremely high light reflectance.

The other use for fillers is in coloring a floor, when a pigmented filler of the desired color tone is applied. This process is identical to bleaching, except the filler is tinted instead of white.

Apply filler with a 4-inch, short-bristle, flat brush. Cover a small area at a time, brushing with the grain first, then across the grain. When it dulls over, but before it hardens, wipe vigorously with burlap or other coarse rags. Wipe across the grain first, then with the grain. Move on to another area until all flooring is filled. Let the filler dry 24 hours and disc sand with fine paper before you apply other finish materials.

GYM AND ROLLER RINK FLOORS

Sanding and other initial preparations for applying a finish to gym floors are identical to those for floors in other installations. Special finishing products are made to provide the gloss and slip- and abrasion-resistance characteristics required, as well as to permit painting of game lines. Use only products manufactured for such applications and follow the specific finishing schedule provided since this varies from one product to another. Deviations from the recommended procedure may effect the quality and performance of the finish.

For those who are installing gymnasium or game floors of hardwood, Appendix A gives the specific markings and measurements for most popular gym games.

PROTECTING THE FINISH

For the final touch of beauty and to protect the finish, apply one or more coats of good wax. Use either a liquid buffing wax/cleaner or paste wax. Use only brands that are designated for hardwood floors, and if a liquid, be sure it has a solvent, not a water, base.

Apply after the finish coat is thoroughly dry and polish with a machine buffer. The wax will give a

lustrous sheen to the floor and form a protective film that prevents dirt from penetrating the finish.

One word of caution: some manufacturers of urethane finishes do not recommend waxing, especially for commercial jobs, because wax may make the floor slippery. Gymnasium and roller rink floors should never be waxed. They require special maintenance products and procedures that are available from several manufacturers who also produce the finishing materials for such installations.

BASE MOLDINGS

Base moldings serve as a finish between the finished floor and wall (Figs. 4-18 and 4-19). They are available in several widths and forms. The simplest is made by butt-jointing two pieces of wood at an inside corner (Fig. 4-20).

A two-piece base consists of a baseboard topped with a small base cap (Fig. 4-21A). When plaster is not straight and true or a hardwood floor needs an expansion gap, the small base molding will more closely conform to the variations than will the wider base alone. A common size for this type of

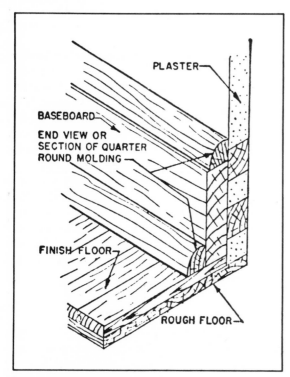

Fig. 4-18. Base moldings serve as a finish between the hardwood flooring and the wall.

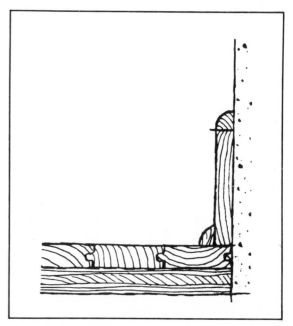

Fig. 4-19. Cross section of molding applied over hardwood flooring.

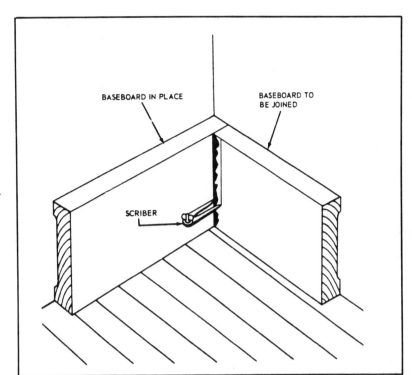

Fig. 4-20. Butt joint baseboard.

baseboard is 5/8 × 3 1/4 inches or wider.

A one-piece base varies in size from 7/16 × 2 1/4 inches to 1/2 × 3 1/4 inches and wider (Fig. 4-21B and 4-21C). Although a wood member is desirable at the junction of the wall and flooring to serve as a protective "bumper," wood trim is sometimes eliminated entirely.

Most baseboards are finished with a base shoe, 1/2 × 3/4 inch in size. A single-base molding without the shoe is sometimes placed at the wall-floor junction.

Square-edged baseboard should be installed with a butt joint at inside corners and a mitered joint at outside corners (Fig. 4-22). It should be nailed

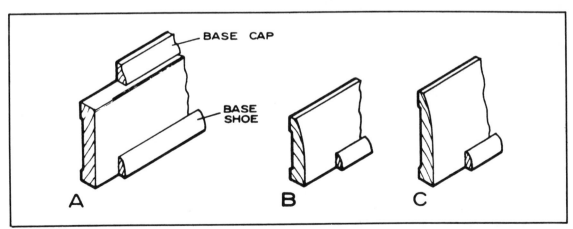

Fig. 4-21. Baseboard topped with a small base cap (A), a narrow ranch base (B), and wide ranch base (C).

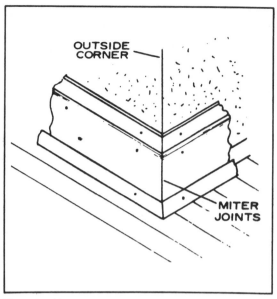

Fig. 4-22. Miter joints at outside corner molding.

to each stud with two 8d finishing nails. Molded single-piece base, base moldings, and base shoe should have a coped joint at inside corners and a mitered joint at outside corners.

A *coped joint* is one in which the first piece is square-cut against the plaster or base, and the second molding coped. It is accomplished by sawing a 45-degree miter cut, and with a coping saw trimming the molding along the inner line of the miter (Fig. 4-23).

The base shoe should be nailed into the subfloor with long, slender nails and not into the baseboard itself or the hardwood flooring. Thus, if there is a small amount of shrinkage of the joists or an expansion of the hardwood flooring, the molding will allow this movement while covering it.

To guide you in the selection and installation of common molding types, let's look at some of the more popular designs (Fig. 4-24).

The term *contour cutting* refers to the cutting or ornamental face curves on stock which is to be used for molding or other trim. Most contour cutting is done on the shaper, equipped with a cutter or blades, or with a combination of cutters and/or blades, arranged to produce the desired contour.

The simple molding shapes, shown in Fig. 4-24 are the quarter round, the half round, the scotia or cove, the cyma recta, and the cyma reversa. The quarter and half round form convex curves; the cove molding forms a concave curve, and the cyma

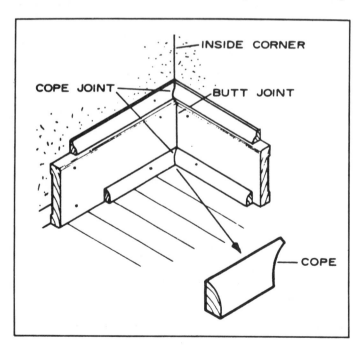

Fig. 4-23. Coping an inside corner molding.

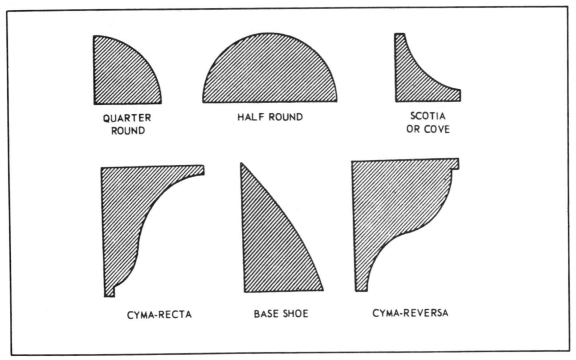

Fig. 4-24. Common molding shapes.

QUARTER ROUND

HALF ROUND

SCOTIA OR COVE

CYMA-RECTA

BASE SHOE

CYMA-REVERSA

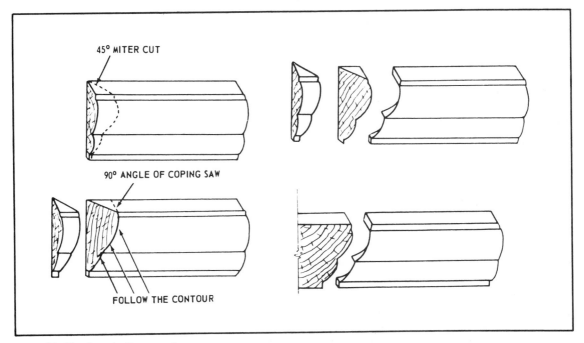

Fig. 4-25. Shaping abutting members.

45° MITER CUT

90° ANGLE OF COPING SAW

FOLLOW THE CONTOUR

moldings are combinations of convex and concave curves.

Inside corner joints between molding trim members are usually made by butting the end of one member against the face of the other. Figure 4-25 shows the method of shaping the end of the abutting member to fit the face of the other member.

First, saw off the end of the abutting member square, as you would for any ordinary butt joint between ordinary flat-faced members. Then miter the end to 45 degrees, as shown in the first and second views of Fig. 4-25. Set the coping saw at the top of the line of the miter cut, hold the saw at 90 degrees to the lengthwise axis of the piece, and saw off the segment shown in the third view, following closely the face line left by the 45-degree miter cut. The end of the abutting member will then match the face of the member, as shown in the third view. This is called a *coping joint*.

You've now completed the finishing of your hardwood floor. The floor has been prepared, sanded, and a finish installed. The molding around the ledge has been cut and installed. All that's left is the enjoying of your beautiful hardwood floor.

To ensure that your floor will give both beauty and function for many years, you'll want to maintain it properly. That's the subject of Chapter 5.

Chapter 5

Maintaining Hardwood Floors

Y OU'VE PUT A GOOD DEAL OF MONEY AND EF-
fort into your beautiful new hardwood floor.
To get the greatest return on your investment, you
must take care of the floor. It is not difficult to do
so. In fact, you'll find that maintaining hardwood
floors is easier than taking care of most other types
of flooring materials.

GENERAL CARE OF HARDWOOD FLOORS

There are a number of things you can do without touching a mop or broom that will give longer
life to your hardwood floor.

Humidity and ventilation are critical considera-
tions for your new wood floor. Relative humidity
of 40 to 50 percent is normally required for long,
trouble-free life. If humidity rises over 50 percent,
prompt air circulation should be initiated by opening interior doors and windows and by activating
the ventilating system. Don't draw warm, moist air
from outdoors, however, because excessive
humidity will cause wood to expand.

Summer months are especially critical. Inspect

your wood floors regularly. If necessary, turn on
the heating system. If less than 35 percent humidity
level persists, use humidification to prevent excessive dryness and possible wood shrinkage. Refer to Fig. 5-1.

When excessive tightening of the floor becomes
noticeable, call your flooring contractor or supplier
immediately. When unusually wide cracks begin to
appear, you should also call your flooring contractor or supplier to solve the problem as quickly as
possible.

Be sure that your air-conditioning system is
operating within the 40- to 50-percent range of normal relative humidity. Ventilation equipment
should be available for year-round use.

Avoid exposure to water from tracking during
periods of inclement weather by protecting your
floors at exterior doorways. Floor protection should
be checked thoroughly to ensure no moisture is
trapped underneath. Windows and doors should be
closed during rainy weather. All leaks must be corrected immediately.

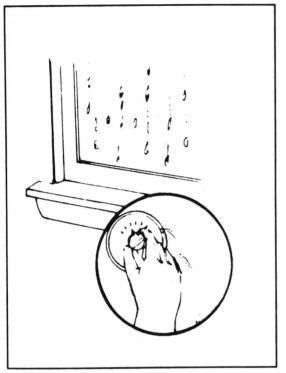

Fig. 5-1. Humidity can be a problem for hardwood flooring (courtesy Maple Flooring Manufacturers Association).

DAILY CARE

Wood floors, properly finished, are the easiest of all floor surfaces to keep clean and look new. Unlike carpeted or resilient floors that show age regardless of care, wood floors can be kept looking like new, year after year, with minimum care.

Both open-grained and close-grained woods are used in flooring. Heading the list of hard, open-grained woods is durable oak, used for an estimated 95 percent of all wood floors. Other hardwoods include northern walnut, pecan, ash, elm, and chestnut. Among the close-grained woods are maple, birch, beech, Douglas fir, and yellow pine. Since the overwhelming majority of wood floors are hardwood, I'll cover the daily care specifically of this type of flooring.

A good rule of thumb for minimum care is to vacuum or dust-mop weekly (Figs. 5-2 through 5-5). An occasional buffing helps remove scuff marks that may appear in the wax coating. Rewax once

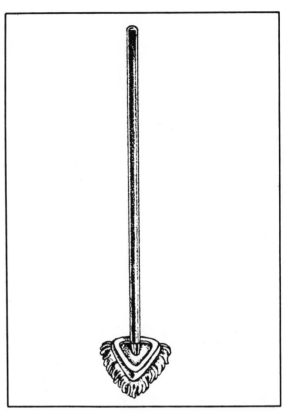

Fig. 5-2. Dust mop (courtesy Maple Flooring Manufacturers Association).

Fig. 5-3. Sweeping the flooring clean (courtesy Maple Flooring Manufacturers Association).

Fig. 5-4. Dust mopping your hardwood floor (courtesy Maple Flooring Manufacturers Association).

Fig. 5-5. Dust mop and vacuum for greatest cleaning (courtesy Harris-Tarkett, Inc.).

or twice a year—or as often as needed in high-traffic areas—using a liquid buffing wax/cleaner combination.

No matter what finish your wood floor has, or what claims the manufacturer makes for his finish, never wash or wet-mop wood floors. Water can seep between boards, leaving dark stains and sometimes warping the boards.

Always read the label of products you use. The recommendations made here on care serve as general guidelines in selecting and using floor maintenance materials. Except for any directions on using water on wood, wherever these guidelines vary from the product's label instructions, always follow the label.

UNDERSTANDING YOUR FINISH

As you learned in the last chapter, there are two principal types of finishes used on wood floors: penetrating seals and surface finishes. Each requires about the same care but when it comes to removing stains or restoring the finish in heavy-traffic areas, methods vary. It helps, therefore, to know what type of finish was used on your floors. If you just put the finish on, as described in Chapter 4, you already know. Maybe you purchased a home with an existing hardwood floor, however. How can

you find out what type of floor finish you have?

As a general rule you can be sure your floor was prefinished at the factory if it has V-shaped grooves along the edges where the boards join and sometimes where the ends butt. Unfinished plank flooring may also have grooved edges, but usually this is a mark of a prefinished floor. Once you have established the floor as prefinished, the flooring manufacturer can tell you whether or not it has a penetrating seal finish. For your convenience, the names and addresses of many primary floor manufacturers are listed in Appendix B of this book.

If the floor has no grooves, it was in all likelihood finished after installation by local craftsmen or a do-it-yourselfer. To determine what kind of finish was used, call the builder, floor finisher, or previous owner if possible. When in doubt, it is safest to assume that a surface finish was used. Treating a penetrating seal finish as though it were a surface finish can do no harm, whereas a surface finish treated as a penetrating seal is likely to be ground away.

MAINTAINING PENETRATING SEALS

A penetrating seal is the finish recommended for most residential floors. As its name implies, the

99

sealer soaks into the wood pores and hardens to seal the floor against dirt and certain stains.

At the surface, it delivers a low-gloss satin finish that wears only as the wood wears. Because of this, color may be added to the liquid sealer at the time of application. The eventual effect of traffic will be far less apparent than with other finishes that only coat the surface. When an area does begin to show wear, it can be refinished easily. The new application will blend into the old without lap marks or other signs of repair.

The beauty and wear resistance of wood floor finishes with a penetrating seal may be further enhanced by wax. A wax coating forms a barrier against the most frequent kind of abrasion and can be easily renewed.

A penetrating seal may be used also as an undercoat for surface finishes, serving as a stain to color the wood before the surface finish is applied. The surface finish used should be compatible with the penetrating seal. An unmatched surface coating may peel.

MAINTAINING SURFACE FINISHES

Surface finishes include polyurethane, varnish, shellac, and lacquer. Maintenance of these finishes is different from that for penetrating seals.

Polyurethane is a blend of synthetic resins, plasticizers, and other film-forming ingredients that produces an extremely durable surface which is moisture-resistant. It is the best choice for a kitchen where the floor is exposed to stains and spills. Polyurethane is available both in high-gloss and matte finishes.

Some manufacturers of polyurethane products say no waxing is required. Many flooring experts, however, believe that you'll get better wear and appearance if you give it the same care as other surface finishes.

Depending on the type used, varnish finishes will be high, medium, or low in gloss. Varnish tends to darken with age and is difficult to touch up. It dries slowly. If the quality is good, varnish will provide a highly durable surface; if not, it tends to become brittle, to powder, and to show white scars.

Shellac is a popular finish for floors in houses built in certain areas of the country. It dries so fast that two coats can easily be applied in one day, and the floor used 8 hours later. Liquid spills, however, can cause hard-to-remove spots on a shellac finish. The abrasive action of footsteps also creates frictional heat that softens the finish and permits entry of dirt. Waxing is essential to protect the finish.

Even faster drying than shellac, lacquer requires real skill in application. It produces a tough, high sheen; but this sheen is difficult to maintain, and scuff marks show easily.

MAINTAINING POLYMER FINISHES

There is a third classification of finishes recently developed that is known as an irradiated polymer. Thus far, it is used primarily in commercial applications. Each brand of flooring using a polymer finish has a different maintenance schedule, available from the manufacturer.

EASY FLOOR CARE

If your floors are new or newly refinished with either a penetrating sealer or a surface finish, start them off right by applying a liquid buffing wax/cleaner or a coating of paste wax. The wax will form a protective barrier for the finish to keep out dirt and potential stain-causing matter so your floors will stay beautiful and resist wear for a long time.

Liquid buffing wax (Fig. 5-6) is easier to use than paste wax and so will probably be used more often. For this reason, many professionals recommend the liquid with these two cautions: the wax (liquid or paste) must be designated for use on hardwood floors, and the liquid must not have a water base. Check the label since some manufacturers recommend their water-based products for wood, while many professionals believe only a solvent-based product should be used. Solvent-based waxes will have the odor of a dry cleaning fluid.

Follow the manufacturer's directions for applying the wax and buff it well, preferably with a 12-inch machine buffer available from rental companies. You may want to buff small areas by hand with clean cloths.

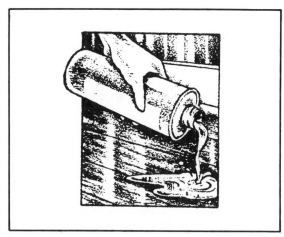

Fig. 5-6. Applying a liquid buffing wax (courtesy National Oak Flooring Manufacturers Association).

Vacuuming is the best way to remove surface dust and dirt before it gets "walked into" the wax and dulls its luster. Vacuuming also pulls accumulated dust from the grooves of prefinished and plank floors. When floor luster has dulled a bit and scuff marks begin to show, you can restore the original beauty often without adding new wax by simply machine or hand buffing.

After 4 to 6 months of wear, inspect your floors closely to see if there's been a dirt build-up, or if the wax has discolored. If your floors were originally finished in a dark tone, you may see a lightening of the finish in traffic areas. If none of

this is apparent, just apply a new coat of wax over the old and buff it well to restore luster. If such conditions do exist, the following procedures can be followed.

Use a combination liquid cleaner/wax (again make sure it has a solvent rather than a water base). For dark floors, choose a buffing wax in a compatible dark color. Spread it with a cloth or fine steel wool. Rub gently to remove grime and the old wax, then wipe clean. Let dry 20 minutes or so and buff. If dull spots remain after buffing, apply a second coat and repeat (Fig. 5-7).

If your floors were stained, it's always a good idea to use a colored wax or cleaner to help maintain the original color. Check the floor care products of your local stores.

MAINTAINING SPECIAL SURFACES

Distressed wood floors have been wire brushed to remove the soft portion of the wood, giving it an antique, textured appearance. The resulting uneven surface tends to trap dirt. Recommended care is regular sweeping with a stiff broom, followed by vacuuming to pick up the loosened dirt.

Such floors are usually stained a dark color to further convey the aged wood effect. What remains after the wire brush treatment, however, are only the toughest wood fibers, and these are somewhat resistant to penetration by the finish color, which means more frequent color renewal. This can be accomplished by using a wax or cleaner/wax combination of the proper color to maintain the original color tone.

REMOVING STAINS

Most stains can be prevented or minimized by keeping the floors well waxed and by wiping up any spilled liquid immediately (Fig. 5-8). Here are some first aid suggestions for common accidents. When removing a stain, always begin at the outer edge and work toward the middle to prevent it from spreading.

Dried milk or food stains. Rub the spot with a damp cloth. Rub dry and rewax.

Fig. 5-7. If dull spots remain, apply a second coat and repeat buffing (courtesy Pennwood Products Co.).

Fig. 5-8. It is important that spilled liquids be wiped up immediately (courtesy Maple Flooring Manufacturers Association).

Stains caused by standing water. Rub the spot with No. 00 steel wool and rewax. If this fails, sand lightly with fine sandpaper. Clean the spot and the surrounding area using No. 1 steel wool and mineral spirits or a proprietary floor cleaner. Let the floor dry. Apply matching finish on the floor, feathering out into the surrounding area. Wax after the finish dries thoroughly.

Dark spots. Clean the spot and surrounding area with No. 1 steel wool and a good floor cleaner or mineral spirits. Thoroughly wash the spotted area with household vinegar (Fig. 5-9). Allow it to remain for 3 to 4 minutes. If the spot remains, sand with fine sandpaper, feathering out 3 to 4 inches into the surrounding area, rewax, and polish.

If repeated applications of vinegar do not remove the spot, apply oxalic acid solution directly onto the spot. Proportions are 1 ounce oxalic acid to 1 quart water, or fractions thereof.

Caution: oxalic acid is a poison; use rubber gloves. Pour a small amount directly on the spot and let the solution stand 1 hour. Sponge the spot with clear water. A second treatment may be helpful if the spot refuses to yield.

If a second application of oxalic acid fails, sand the area with No. 00 sandpaper and apply matching finish, feathering out into the surrounding floor

area. Let dry. Buff lightly with No. 00 steel wool. Apply second coat of finish, let dry, and wax. If the spot is still visible, the only remaining remedy is to replace the affected flooring. Note: oxalic acid is a bleaching agent. Whenever it is used, the treated floor area will probably have to be stained and refinished to match the original color.

Heel marks, caster marks, etc. Rub vigorously with fine steel wool and a good floor cleaner. Wipe dry and polish.

Ink stains. Follow the same procedures as for other dark spots.

Animal and diaper stains. Spots that are not too old may sometimes be removed in the same manner as other dark spots. If spots resist cleaning efforts, the affected flooring can be refinished.

Mold. Mold or mildew is a surface condition caused by damp, stagnant air. After seeing that proper ventilation is provided for the room, the mold can usually be removed with a good cleaning fluid.

Chewing gum, crayon, candle wax. Apply ice until the deposit is brittle enough to crumble off. Cleaning fluid poured around the area (not on it) soaks under the deposit and loosens it.

Cigarette burns. If not too deep, steel wool will often remove them. Moisten steel wool with soap and water to increase effectiveness.

Fig. 5-9. Thoroughly wash spotted area with household vinegar (courtesy National Oak Flooring Manufacturers Association).

Alcohol spot. Rub with liquid or paste wax, silver polish, boiled linseed oil, or cloth barely dampened in ammonia. Rewax the affected area.

Oil and grease stains. Rub on a kitchen soap having a high lye content, or saturate cotton with hydrogen peroxide and place over the stain. Saturate a second layer of cotton with ammonia and place over the first. Repeat until the stain is removed.

Wax build-up. Floors that have not had proper care may acquire wax build-up. Strip all the old wax away with mineral spirits or naptha. (Caution: naptha is extremely flammable. Use only where there is no open flame or danger of spark and provide ample ventilation.) Use cloths and fine steel wool and remove all residue before applying new wax. It's a good idea to perform this complete stripping job every now and then instead of using the liquid cleaner/wax process. Stripping removes all the old wax and dirt, which build up inevitably over a period of time and partially hide the beauty and color of the wood grain.

REPAIRING THE FINISH

Small areas of floors finished with penetrating seal can be repaired successfully without professional help. With special care and skill, you may also be able to repair varnish and polyurethane finishes yourself. Such repair may be necessary after stain removal or water damage. Use steel wool to smooth out the affected boards and an inch or two of the surrounding area. Then brush on one or more thin coats of finish, feathering it into the old finish to prevent lap marks. Allow plenty of drying time between coats and then wax well.

One caution: don't attempt this if you have a lacquer or shellac finish since these are almost impossible to patch successfully. For a small, relatively inconspicuous area, you might get by with steel wool cleaning followed by paste wax. You won't get an exact match, but it could serve as temporary repair. The alternate is sanding to expose bare wood over the entire room and applying new finish. Such refinishing will be covered in detail in Chapter 6.

SOLVING CRACKS AND SQUEAKS

All the wood in your home will contract or expand according to the moisture in the air. Doors and windows may swell and stick during rainy seasons. In dry, cold weather, cracks and fine lines of separation may appear in wall cabinets and furniture. This is characteristic of wood because wood is a product of Nature, and its natural quality is what makes it desirable.

The same reaction to humidity or the lack of it is happening constantly in your wood floors. Tiny cracks between edges of boards may appear when unusually dry conditions are produced by your heating system. This condition can usually be corrected simply by installing a humidifier. With the proper balance of moisture content in the house, both family and floors benefit from a healthier environment. See Fig. 5-10.

When interiors become damp in rainy weather, boards may expand so that edges rub together and produce a squeak. Improper fastening of the floor or subfloor can also cause squeaks.

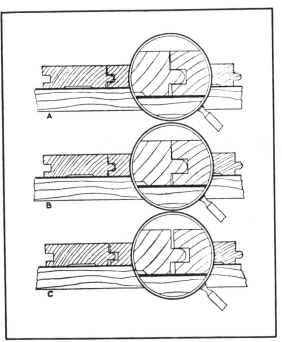

Fig. 5-10. How moisture content affects expansion and contraction, and thus cracks, in hardwood flooring.

Fig. 5-11. Products for fixing flooring squeaks: talcum powder, graphite, and wax (courtesy National Oak Flooring Manufacturers Association).

To correct this situation, first try lubrication (Fig. 5-11). A liberal amount of liquid wax may do the job, or sift a small amount of powdered soap stone, talcum powder, or powdered graphite between adjacent boards where the noise occurs. Another method is to drive triangular glazier points between the strips using a putty knife to set them below the surface.

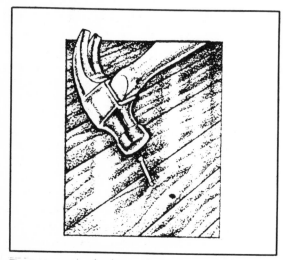

Fig. 5-12. Drive finishing nail through pilot holes to nail down a squeak (courtesy National Oak Floor Manufacturers Association).

If that method doesn't work, drive 2-inch finishing nails through pilot holes drilled into the face of the flooring (Fig. 5-12). Nails should go through both edges of the boards. Set them with a nail punch and hide with matching color putty and wax.

If these efforts don't solve problems with squeaks and cracks, repairs must be made. Let's consider how the repairs for these and other hardwood floor problems can be accomplished by the do-it-yourselfer.

FLOOR REPAIRS

As you learned, the floors in most homes are composed of two separate layers. The bottom layer is called the subflooring and is made of rough tongued-and-grooved lumber nailed directly to the floor joists. In some cases, the subflooring will run diagonally to the joists, while in others it will run at right angles to them. A layer of building paper covers the subfloor to keep out dust and dirt, and the finish or hardwood floor is laid over this paper. The finish floor runs at right angles to the subfloor and is nailed to it. The finish floor can be either of planks, blocks, or strips of hardwood.

Repairing Creaking Floors

In most cases, a creaking floor is caused by a loosening of the nails holding the subfloor to the joists. The nails may either pull loose or be loosened by shrinkage in the wood. Creaking is usually in the subfloor, but will sometimes occur in the finish floor, particularly if the floor was put down before the wood was completely seasoned.

If the creak is in the subflooring and the underside is exposed, as when the flooring functions as the ceiling for an unfinished basement, drive a small wedge between the joist and the loose board (Fig. 5-13). The wedge will take up the play in the board and the noise will stop. If several boards are loose, nail a piece of wood to the joist high enough to prevent these boards from moving down (Fig. 5-14). The nailheads will keep the boards from moving up, effectively ending the noise.

In many cases, it is impossible to reach the

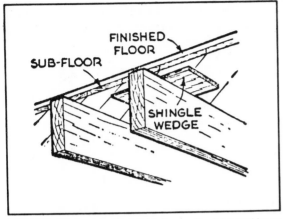

Fig. 5-13. Drive a small wedge between the joist and loose board to stop a creaking floor.

subflooring without tearing up the finish floor or moving a ceiling. Since neither of these alternatives is feasible, the only alternative is to try to locate the floor joist by tapping on the floor. If a floor joist near the creak can be located, then 2- or 3-inch finishing nails can be driven through the finish floor and subfloor into the joist (Fig. 5-15). Drive the nail at an angle, and when it is near the surface of the floor, use a nail set to drive it below the surface of the wood to prevent hitting the finish floor with a hammer and marring the finish. Make the nail inconspicuous by filling the hole with putty or stain to match the rest of the floor.

Occasionally a creaking floor will be caused by a loose board in the finish floor. This board can be located by its movement when weight is placed upon it. Use 2-inch finishing nails and drive them

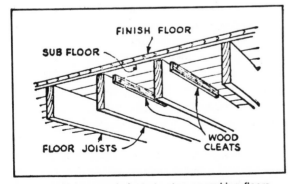

Fig. 5-14. Using wood cleats to stop squeaking floors.

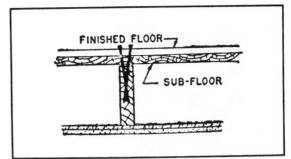

Fig. 5-15. Stop squeak by driving finish nails into joists.

in at an angle, using the nail set in the manner just described.

Sometimes boards in the finish floor warp to such an extent that they pull away from the subfloor and bulge. They can be driven back into place by putting a piece of heavy paper and a block of wood over them and tapping the wood sharply with a hammer. The piece of wood prevents the hammer from damaging the floor. Take care in doing this, however, for the thin edges of tongued-and-grooved boards can easily be split.

Repairing Sagging Floors

When a sagging floor is found in a very old house it is generally because the floor joists and girders have been weakened by rot or insects. In a new house, a weak floor can, in most cases, be blamed on the builder. A flooring built of under-sized materials and tacked together will be neither substantial nor capable of bearing much weight.

In dealing with a weak and sagging floor, you will first have to raise it to its proper level. If it is the first floor, with a basement underneath, the work is in the range of the do-it-yourselfer. Use heavy lumber and a screw jack to accomplish the work (Fig. 5-16). The size of the lumber should be about 4 × 4 inches. Place one of the 4- × -4 timbers on the basement floor directly under the sag and put the screw jack on top of it. This beam will distribute the weight of the flooring over a relatively large portion of the basement flooring. If the basement flooring is of heavy concrete, this step will not be required. Nail a piece of 4- × -4 along the sagging joists. Use a third piece of timber as a vertical

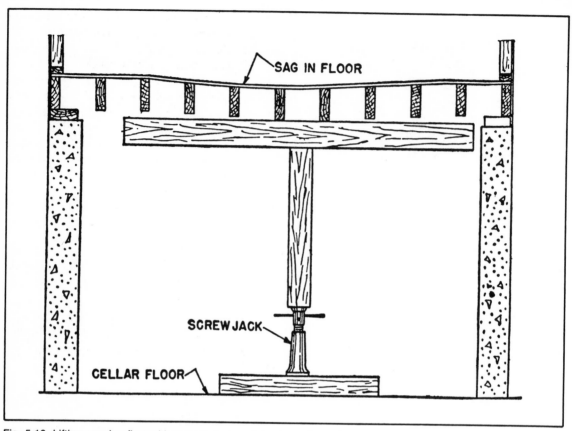

Fig. 5-16. Lifting sagging floor with a screw jack and 4-x-4 wood members.

beam from the top of the jack to the under portion of the 4- × -4 nailed to the joists. Turn up the jack until the floor is level.

Don't attempt to bring the floor to a level position all at once. If you do so, you are almost sure to crack the plaster walls and ceiling in the room above. Raise the jack only a fraction each week, and you will avoid doing extensive damage to the room above. Check the position of the floor with a level, and when it is correct, measure the distance from the bottom of the horizontal 4- × -4 to the floor of the basement. Cut a piece of 4- × -4 to this length. Turn the jack up enough to allow this beam to stand on end under the horizontal 4- × -4. Make sure that it is perfectly vertical and that it rests firmly on the floor. Remove the jack, along with the other timbers, leaving only the one vertical and one horizontal 4- × -4.

If one entire floor is sagging, it will probably be necessary to use more than one vertical support. In this case, place a vertical 4- × -4 under each end of the horizontal beam.

Another means of raising a floor is to use metal posts with screw jacks built into them (Fig. 5-17). The post is provided with two plates, one of which rests on the basement floor, while the other fits between the top of the post and the joist or girder to be raised. These posts are made so that they can be adjusted to different heights. Once the floor has been brought to the right level, the posts can be left as a permanent support. As before, turn the jack only a small amount each week so that the floor will be raised slowly.

When part of the total weight of a floor and the objects on it is supported by posts, it is important that each post have the proper footing. Most con-

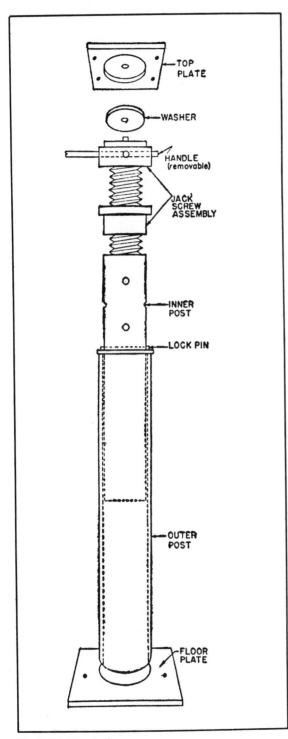

TOP PLATE

WASHER

HANDLE (removable)

JACK SCREW ASSEMBLY

INNER POST

LOCK PIN

OUTER POST

FLOOR PLATE

Fig. 5-17. Floor jack.

crete floors in the basement are rather thin, and it is often necessary to prepare the floor before installing the posts.

To make a substantial footing for the posts in the basement floor, break up about 2 square feet of the concrete floor at the point where the post is to stand. Do this work with a heavy hammer (don't ruin a good claw hammer on it), or with a piece of pipe. Once the surface is broken, dig a hole about 12 inches deep and fill it with concrete made with 1 part cement, 2 parts sand, and 3 parts coarse aggregate. Level this with the floor, making a smooth surface, and allow about 1 week for the footing to dry before placing the posts upon it. Cover the concrete during this period and keep it moist.

Fortunately, most defects are associated with the first floor, and the basement or crawl space underneath allows the installation of posts and other kinds of reinforcement. Sagging floors above the first floor level cannot be practically remedied, short of taking up the flooring and making extensive repairs. For this situation, it's best to call a good carpenter to do the work.

Repairing Floor Cracks

If the wood in a matched hardwood floor is properly seasoned, there should be very few cracks appearing between the boards. Many houses, however, are equipped with plank floors in which cracks of varying size are almost sure to appear between each board. In very old houses these cracks can be quite large. There are several kinds of plastic fillers, but many of them tend to shrink and crack as they dry. A good filler can be made of sawdust and wood glue mixed into a paste. If possible, the sawdust should be of the same wood as the flooring.

Before you attempt to fill a crack, clean it out, because any dirt in it will prevent the filler from adhering to the wood. Pack the filler in tightly so that it stands slightly higher than the surface of the floor. After it is dry, sand the top to the floor level and apply a little stain to match the finish on the rest of the floor. Over very wide cracks, glue a thin strip of wood and sand or plane it to match the floor surface (Fig. 5-18).

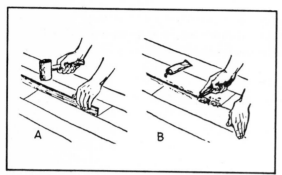

Fig. 5-18. Filling flooring cracks with a wood strip (A) and plastic crack filler (B).

Replacing Flooring Boards

Wooden floors become uneven through much wear and leave high places, particularly where knots and heads of nails occur, since these possess a greater resistance to wear. Unevenness occasionally takes place along the edges of the floor boards, which become raised due to the boards curling up as they warp. If the underlaying joists warp and twist, or if there is some settlement of the foundation walls on which these joists rest, the level of the floor may be disturbed.

To replace boards, first study the diagram of a single wooden plank floor given in Fig. 5-19. The joists are usually about 2 inches wide. The boards, which cross the joists at right angles, are nailed at the centerline of the joist. If the ends of two boards meet in the form of a heading joint on one joist, both boards must be nailed to the joist. To remove the boards without damage is not easy. If one board is to be discarded, you can bore a round hole as near as convenient to the side of the joist and use a keyhole saw and compass saw to cut one board close up to the joists. If the other end of the board runs to a heading joint on another joist, work back along the board, prying it up at the intervening joist and taking it off the joist where it ends.

Perhaps only part of the lifted board is defective, in which case you can cut it across to end on a suitable joist, ready for replacement later. In prying up the board, the nails will most likely be pulled up out of the joist. Rest the board, bottom side up, on a stool or saw horse and tap the nails back sufficiently for the heads to be gripped by pincers or by the claws of the hammer. Obstinate boards may have to have the nails punched fully into the joist to free the board.

If a heading joint is not conveniently near, the

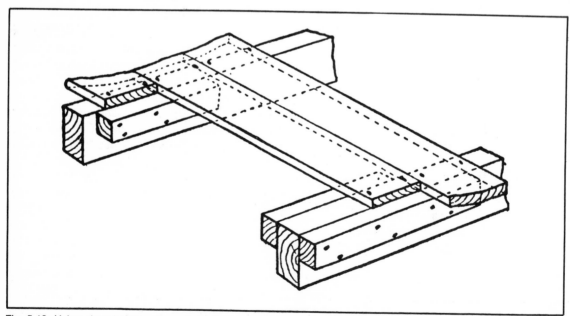

Fig. 5-19. Using short lengths of wood attached to the side of a joist to repair defective floorboards.

board may have to be cut through at two places to remove the defective part. Bore a 1/2-inch hole and cut across the board using a keyhole saw.

Find the run of the joists, indicated by the position of the nailheads. On the assumption that they mark approximately the middle of the joist, measure 1 inch to either side of the nail, then square a line across with a square. Mark a pencilled line. Put a fine bradawl through the board about 1/4 inch away from the pencil line, on the free side, then bore a hole. If these dimensions have been correctly established, the joist will be visible, and the saw can be put through the cut alongside the joist, across the board. As soon as the cut is long enough, take out the keyhole saw and insert a compass saw or a small crosscut saw, and complete the cut.

Beware of water pipes, gas lines, and electric wiring when cutting the boards. They run usually in the space between joists. When there is a room below the floor where the work is in progress, some guide to the position of pipes and cables can be estimated from the location of lighting and plumbing fixtures in that room.

Having cut the board, the ends of the fixed parts can be trimmed with a sharp chisel to a square edge. If several boards have to be cut away, take them back to joists one or two away to the right or left from the one originally selected for the patch. In other words, break the joints so that a board extending over a given joist is next to one at which a board ends, and so on. Thus you will not get a weak line of joists running along the same joist.

A typical job is shown in Fig. 5-20. The joists are numbered and the floor boards lettered. A heading joint is shown at XX on board B. It is not always practical to make heading joints when replacing the boards, and the best thing to do is to support the ends of the replacement boards by nailing or screwing a strong cleat to the side of the joist where the end of the new board will rest. The cleat should be at least 1 1/2 inches thick and about 3 inches wide. Take it halfway along and under the boards adjoining the one that it will support. This arrangement is illustrated in Fig. 5-19.

Two boards can be cut through obliquely when they have to be jointed over a single joist. In this instance it is assumed that both boards can be taken to the bench and cut by a tenon saw to a suitable angle. This makes a neat and sound job, with the nails being driven through the oblique portion. The angle can be marked across the edges of the board with an adjustable level.

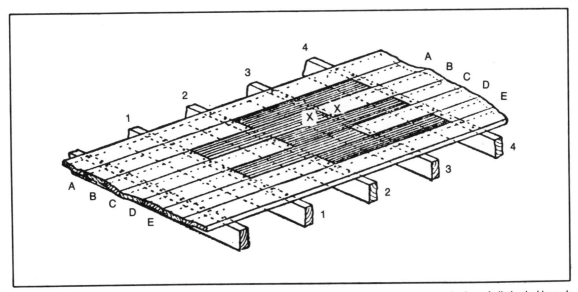

Fig. 5-20. Replacing defective floorboards between joists 2 and 3 requires the removing and replacing of all shaded boards in this typical example.

Repairing Block and Parquet Floors

Loose blocks in a floor, if there are not many, should be removed. It will then be possible to scrape off the old mastic underneath. Put in fresh mastic, which you can get at hardware stores, and bed in the block. If the defect is extensive, the repairs may be more than an amateur can successfully undertake.

Parquet floors are glued and usually screwed into place. Dampness may cause the parts of the design to come loose. In such cases the cause of the dampness should be found before you attempt a remedy. Wood glue can be used to hold the different parts of a pattern together if a whole unit is defective. These diamonds and other designs are bedded upon a piece of low-quality material with an open weave which helps to hold them together. It will probably be best to unite the various parts for an element first, then let the glue harden before relaying to ensure that all joints are firmly set and safe to handle.

Those are the basics of maintaining and repairing new and old hardwood floors. You can see that while repairs can be made by the do-it-yourselfer, they are not easy. Therefore, the best insurance against major flooring repair is good maintenance methods performed on efficiently selected flooring.

You may someday need to refinish your new or old hardwood floor, however. Complete instructions on how to do so are presented in Chapter 6.

Chapter 6

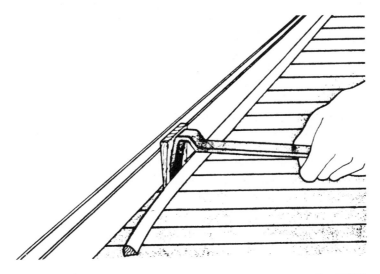

Refinishing Hardwood Floors

THERE HAS BEEN A REVIVAL OF INTEREST IN older homes. Many people are spending their weekends replacing molding, stripping floors, insulating walls, refinishing woodwork, and revitalizing the works of past craftsmen. Others are remodeling older homes for economy and function, finding that the "old ways" are often the best.

In this final chapter you'll learn how to refinish a hardwood floor. In most cases, it can be easily done by the average do-it-yourselfer with an understanding of the process and some rented equipment. Take your time. Learn how it's done; talk with your supplier; talk to others who have refinished hardwood floors. You'll find a wealth of information and opinions on how best to renew hardwood flooring.

If you haven't already, please read Chapter 4. It discusses the types of finishes commonly used, equipment, how the finishes are applied, and other background information vital to the flooring restorer.

WHEN TO REFINISH

Where floors have become badly discolored and worn by neglect or improper maintenance, the most practical procedure, and often the only one that will restore a fine finish, is to have the old finish removed and the floor reconditioned by power sanding (Fig. 6-1). Where the floors have been reasonably well-maintained but the finish has become dingy with age, refinishing without power sanding may be practical. The method of removal of the old finish depends upon the kind of finish that was originally installed.

Varnish Finish

Old, discolored varnish is usually removed most easily by power sanding. If desired, this can be done with liquid varnish remover. Alkaline solutions in water and removers sold in powder form to be dissolved in water should not be used.

Follow the directions for using liquid remover

carefully. Since some of the old, discolored varnish remains imbedded in the wood, do not expect complete restoration of the natural wood. Clean traffic channels where the old varnish has worn through and where dirt has been ground into the wood thoroughly by sanding.

Shellac Finished

Old shellac and wax finishes that have merely become soiled by dirt in the wax coating may be cleaned by going over the floor with steel wool saturated with clean turpentine. Any white spots in the shellac caused by contact with water may be taken out by rubbing lightly with a soft cloth moistened with denatured alcohol. The alcohol must be used with care, however, to avoid cutting the shellac coating.

On floors where the dirt is ground into the shellac itself or white spots penetrate through the coating, more drastic treatment is necessary. First, wash the floor with a neutral or mildly alkaline soap solution. Then scour the floor with No. 3 steel wool and denatured alcohol. If the floor boards are level and are not warped or cupped, the scouring can be done to advantage with a floor polishing machine fitted with a wire brush to which a pad of the No. 3 steel wool is attached. After scouring, the floor should be wiped clean and allowed to dry thoroughly before you refinish with shellac or other finish.

Oil Finish

Some older floors have been finished with linseed oil. To refinish, clean the surface and apply a new coating of oil. Make needed repairs and smooth rough spots if the floor is in poor condition.

To remove oil, some professionals recommend that you wet about 10 square feet of floor with a mop and warm water and sprinkle the area liberally with a mixture of one part soap powder and three parts trisodium phosphate. Scrub the floor with a stiff brush, using only as much water as needed to form an emulsion and float the oil to the surface. As the oil is loosened, remove it with a squeegee and mop. Rinse the area and mop dry. Treat the

other sections in the same way. When the entire floor has been cleaned, let the surface dry at least 24 hours, then sand with a machine and finish as desired.

A reminder: many professional floor refinishers recommend that no water touch the wood since it is quickly absorbed and may cause cracking and swelling of the wood flooring once the finish is installed. Discuss this with your floor supplier.

REFINISHING

Those are some of the ways that you can refinish your hardwood floor with little or no need for a floor sander. In many cases however, it is best to roll up your sleeves and plan on renting a floor sander for a day or two in order to completely strip an old floor. Table 6-1 suggests the selection of sandpaper you should buy for refinishing an old floor.

First, remove all furniture, rugs, and draperies from the room. If you're planning on painting or applying wallcovering, do that work before you refinish the floor so paint or paste won't drip on the new floor. Vacuum the floor.

Look for protruding nailheads and drive them down with a nail set. Tighten any loose boards by face-nailing, preferably into joists, then countersink with a nail set.

Before sanding an old floor, remove the base shoe (Fig. 6-2). Use a wood wedge behind the pry bar to protect the baseboard from damage by the bar. Figure 6-3 illustrates the cross section of a typical hardwood block floor.

Table 6-1. Sanding Old Floors.

FLOOR Covered with Varnish, Shellac, Paint, etc.	OPERATION	TYPE OF PAPER
	First Cut	Coarse 3½ (20)
	Second Cut	Medium 1½ (40)
	Finish Sanding	Fine 2/0 (100)

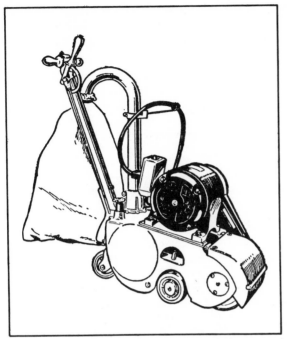

Fig. 6-1. An electric floor sander can be rented to refinish an old hardwood floor.

SANDING AN OLD FLOOR

Most oak and many other types of flooring can be sanded and refinished a number of times. Thinner floors should be refinished with caution because repeated sanding could wear through to the subfloor. To determine the floor thickness, remove a floor heating register or the shoe mold and baseboard (Fig. 6-3) so that an edge of the flooring is exposed and can be measured.

When you are refinishing thinner floors—1/2 or 3/8 inch—use the floor polishing machine and screen abrasive rather than the drum sander. Remove as little of the surface as possible. Do not do this for facenailed 5/16-inch square-edge, or 5/16-inch square-edge installed in mastic, however.

The 5/16-inch type is almost always face-nailed; others are blind-nailed through the tongue. The nails must first be set (driven deeper into the wood) to permit sanding the floor.

The following instructions apply to standard 3/4-inch strip, plank, and block floors and, with the cautions just mentioned, to the thinner materials. Use an open-face paper to remove the finish. The heat and abrasion of the sanding operation make the old finish gummy and will quickly clog normal sanding paper. When you get down to the new wood, you can switch to regular paper for the final finishing cuts.

The number of cuts required for the refinishing operation will be largely determined by the condition of the old floor and the thickness of the finish being removed. If the surface is in good shape and

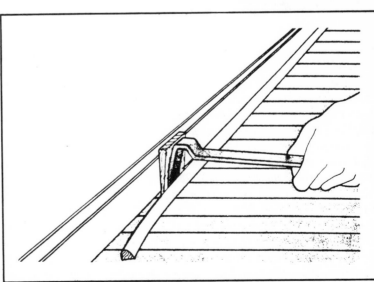

Fig. 6-2. Before sanding an old floor, remove the base shoe (courtesy National Oak Flooring Manufacturers Association).

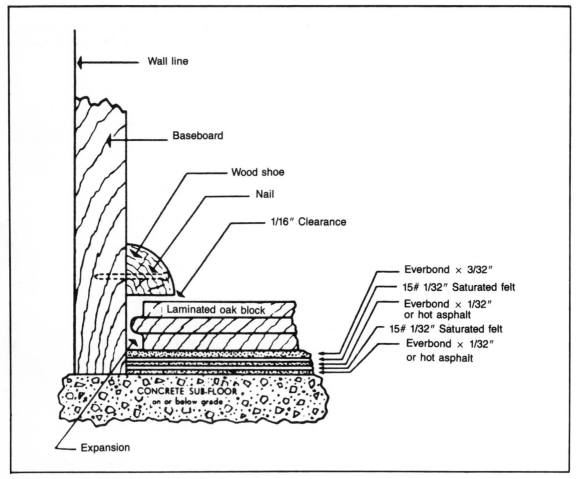

Fig. 6-3. Checking the thickness of the flooring.

Image labels: Wall line, Baseboard, Wood shoe, Nail, 1/16" Clearance, Everbond × 3/32", 15# 1/32" Saturated felt, Everbond × 1/32" or hot asphalt, 15# 1/32" Saturated felt, Everbond × 1/32" or hot asphalt, Laminated oak block, CONCRETE SUB-FLOOR on or below grade, Expansion

has no thick built-up of old finish and wax, one pass with the disc sander and extra-fine paper may be sufficient. Just be sure you've removed all the old finish.

If the floor is badly scarred or boards are cupped dished, use as many cuts as are necessary to get a smooth, unblemished surface. Make the first one or two cuts at a 45-degree angle with medium-grit paper. Follow the instructions given in Chapter 4 for sanding a new floor on the succeeding cuts.

NEW FINISHES

As you learned in Chapter 4, there are two basic finishes applied to hardwood floors: penetrating seals and surface finishes. Most manufacturers of penetrating seal finishes also make a renovator or reconditioning product for use when traffic or other conditions cause discoloration or wear of the finish. These products restore the floor to its original appearance without the need for sanding. They can be used on most prefinished flooring, which may be identified by the beveled edge on each strip or block.

Another way to determine if a floor was originally finished with a penetrating seal is to scratch the surface in a corner or some other inconspicuous space with a coin or other sharp-edged object. If the finish does not flake off, a penetrating seal was probably used, and a renovator product can be applied to restore its original beauty.

If the condition of the flooring is very bad, sanding and application of a complete new finish may be needed. Once the floor is completely stripped and cleaned, follow the directions in Chapter 4 for finishing hardwood floors.

With most surface finishes, the recommended method of restoration is to sand off the old finish and apply a completely new finish, using penetrating seal or polyurethane. If the floor was originally finished with polyurethane and is in good condition, it should be cleaned and screen-disced to rough up the old finish.

A word of caution: adhesion of polyurethanes, epoxies, and urea-formaldehyde finishes is affected by wax and grease, as well as some types of stains, bleaches, or sealers. Further more, one type of polyurethane may not be compatible with another

type. Table 6-2 offers information on the type of wood type of grain, and notes on finishing that may guide you in refinishing your old hardwood floor.

REFINISH PLANING

Sometimes you will run across a strip or block of hardwood flooring that needs to have the surface planed before it can be refinished. Let's consider how this is done.

The plane is the most extensively used hand shaving tools. The large family of planes includes bench planes and block planes, designed for general surface smoothing and squaring, and other planes designed for special types of surface work.

The principal parts of a bench plane and the manner in which they are assembled are shown in Fig. 6-4. The part at the rear that you grasp to push

Table 6-2. Wood and Wood Finishes.

| | Type .... | Soft | Hard | | Notes on Finishing |
Name of Wood	Grain	Closed	Open	Closed	
Ash		......	X	......	Requires filler.
Alder		X	......	......	Stains well.
Aspen		......	......	X	Paints well.
Basswood		......	......	X	Paints well.
Beech		......	......	X	Paints poorly; varnishes well.
Birch		......	......	X	Paints and varnishes well.
Cedar		X	......	......	Paints and varnishes well.
Cherry		......	......	X	Varnishes well.
Chestnut		......	X	......	Requires filler; paints poorly.
Cottonwood		......	......	X	Paints well.
Cypress		......	......	X	Paints and varnishes well.
Elm		......	X	......	Requires filler; paints poorly.
Fir		X	......	......	Paints poorly.
Gum		......	......	X	Varnishes well.
Hemlock		X	......	......	Paints fairly well.
Hickory		......	X	......	Requires filler.
Mahogany		......	X	......	Requires filler.
Maple		......	......	X	Varnishes well.
Oak		......	X	......	Requires filler.
Pine		X	......	......	Variable depending on grain.
Teak		......	X	......	Requires filler.
Walnut		......	X	......	Requires filler.
Redwood		X	......	......	Paints well.

Note: Any type finish may be applied unless otherwise specified.

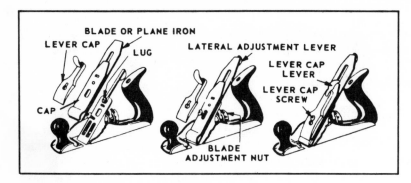

Fig. 6-4. Components of a typical bench plane.

the plane ahead is called the *handle*; the part at the front that you grasp to guide the plane along its course is called the *knob*. The main body of the plane, consisting of the bottom, the sides, and the sloping part which carries the plane iron, is called

There are three types of bench planes: the jointer plane or fore plane (Fig. 6-5), the jack plane (Fig. 6-6), and the smooth plane (Fig. 6-7). All are used primarily for shaving and smoothing with the grain; the chief difference is the length of the sole.

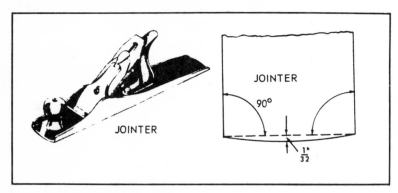

Fig. 6-5. Jointer or fore plane.

the *frame*. The bottom of the frame is called the *sole* and the opening in the sole, through which the blade emerges, is called the *mouth*. The front end of the sole is called the *toe*, and the rear end is the *heel*.

The sole of the smooth plane is about 9 inches long; the sole of the jack plane about 14 inches long, and the sole of the jointer plane from 20 to 24 inches long. The longer the sole of the plane, the more uniformly flat and true the planed surface will be.

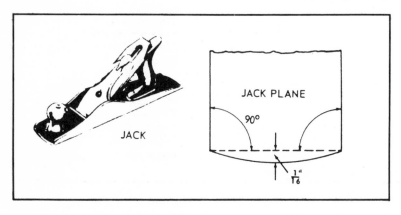

Fig. 6-6. Jack plane.

Fig. 6-7. Smoothing plane.

Consequently, the bench plane you should use depends upon the requirements with regard to surface trueness.

The *smooth plane* is, in general, smoother only; it will plane a smooth, but not an especially true,

plane a truer surface than the smooth plane. It is more often used for planing larger strip or block wood flooring, in or out of the floor.

The *jointer plane* is used when the planed surface must meet the highest requirements with re-

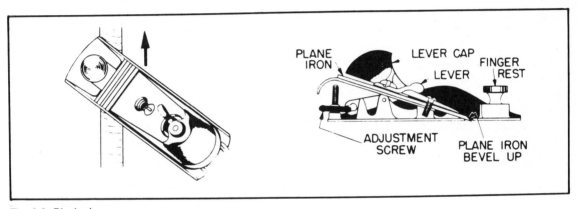

Fig. 6-8. Block plane.

surface in a short time. It is also used for cross-grain smoothing and squaring of end stock. It is especially useful for planing smaller wood strip flooring.

The *jack plane* is the "jack-of-all-work" of the bench plane group. It can take a deeper cut and

gard to trueness. A jointer plane would be used on the floor to level adjoining hardwood strips or blocks.

A *block plane* and the names of its parts are shown in Fig. 6-8. Note that the plane iron in a block

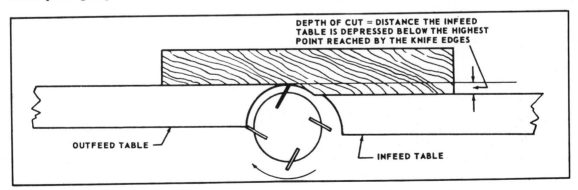

Fig. 6-9. Jointer operation.

plane doesn't have a plane iron cap. Also unlike the iron in a bench plane, the iron in a block plane goes in bevel-up. The block plane, which is usually held at an angle in the work, is used chiefly for cross-grain squaring of end stock. It is also useful for smoothing all plane surfaces on very small work.

Another shaving tool that you may be able to use for smoothing a hardwood flooring strip or block is the *jointer* (Fig. 6-9). The jointer is a machine for power-planing stock on faces, edges, and ends. The planing is done by a revolving cutterhead equipped with two or more knives. The table consists of two parts on either side of the cutterhead. the stock is started on the infeed table and fed past the cutterhead onto the outfeed table. The surface of the outfeed table must be exactly level with the highest point reached by the knife edges. The surface of the infeed table is depressed below the surface of the outfeed table an amount equal to the desired depth of cut. A jointer could be used to shave the surface off new hardwood boards that replace old ones or to lower the depth of all hardwood floor boards in a room prior to installation.

REFINISH CHISELING

As you refinish a hardwood floor or remodel to accommodate new heating vents in the floor, new walls, piping, or cabinets, you may need to change the shape of the hardwood flooring being repaired or installed. The common wood chisel is extremely useful for this task.

A wood chisel should always be held with the flat side, or back side, against the work for smoothing and finishing cuts. Whenever possible, it should not be pushed straight through an opening, but should be moved laterally at the same time that it is pushed forward. This method ensures a shearing cut, which with care will produce a smooth and even surface work even when the work is cross-grain. On rough work, use a hammer or mallet to drive the socket-type chisel.

On fine work, use your hand as the driving power on tang-type chisels. For rough cuts, the bevel edge of the chisel is held against the work. Whenever possible other tools, such as saws and planes, should be used to remove as much of the

waste as possible and the chisel used for finishing purposes only.

To chisel horizontally with the grain, grasp the chisel handle in one hand with your thumb extended towards the blade (Fig. 6-10). The cut is controlled by holding the blade firmly with the other hand, knuckles up, and your hand well back of the cutting edge. The hand on the chisel handle is used to force the chisel into the wood. The other hand pressing downward on the chisel blade regulates the length and depth of the cut.

To chisel horizontally across the grain, hold the work so that it does not move. Remove most of the waste wood by using the chisel with the bevel down. On light work, use hand pressure or light blows on the end of the chisel handle with the palm of your right hand. On heavy work, use a mallet. To avoid splitting at the edges, cut from each edge to the center and slightly upward so that the waste wood at the center is removed last (Fig. 6-10).

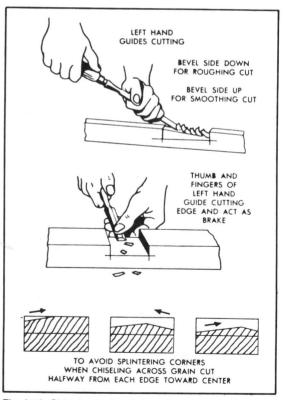

LEFT HAND
GUIDES CUTTING

BEVEL SIDE DOWN
FOR ROUGHING CUT

BEVEL SIDE UP
FOR SMOOTHING CUT

THUMB AND
FINGERS OF
LEFT HAND
GUIDE CUTTING
EDGE AND ACT AS
BRAKE

TO AVOID SPLINTERING CORNERS
WHEN CHISELING ACROSS GRAIN CUT
HALFWAY FROM EACH EDGE TOWARD CENTER

Fig. 6-10. Chiseling horizontally with and across the grain.

Make finishing cuts with the flat side of the chisel down. Never use a mallet when making finishing cuts, even on large work. One-hand pressure is all that is necessary to drive the chisel, which is guided by the thumb and forefinger of your other hand. Finish cuts should also be made form each edge toward the center. Do not cut all the way across from one edge to the other, or the far edge may split.

To cut a round corner on the end of a piece of hardwood flooring, first lay out the work and remove as much waste as possible with a saw (Fig. 6-11). Use the chisel with the bevel side down to make a series of straight cuts tangent to the curve. Move the chisel sideways across the work as it is moved forward. Finish the curve by paring with the level side up. Convex curves are cut in the same manner as round corners.

When cutting a concave curve with a chisel, remove most of the waste wood with a coping saw or a compass saw. Smooth and finish the curve by chiseling (Fig. 6-11) with the grain, holding the chisel with the bevel side down. Use one hand to hold the chisel against the work. Press down on the chisel with the other hand and, at the same time, draw back on the handle to drive the cutting edge in a sweeping curve. Care must be used to take only light cuts, or the work may become damaged.

Vertical chiseling (Fig. 6-12) means cutting at right angles to the surface of the wood, which is horizontal. This technique is used to make a hole in a hardwood flooring strip or block to allow for the passage of a cable, pipe, or other element. Usually it involves cutting across the wood fiber. When you are chiseling vertically across the grain, use a mallet to drive the chisel. A mallet is usually necessary when you are chiseling hardwood. Use a shearing cut in cutting across the grain.

To keep you safe during chiseling, here are a few reminders:

☐ Secure the work so that it cannot move.

☐ Keep both hands back of the cutting edge at all times.

☐ Don't start a cut on a guideline. Start slightly away from it so that there is a small amount of material to be removed by the refinishing cuts.

Fig. 6-11. Chiseling corners and curves on hardwood flooring.

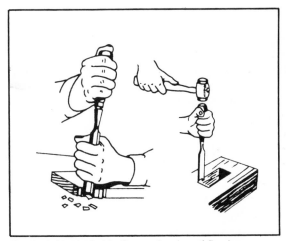

Fig. 6-12. Vertical chiseling on hardwood flooring.

☐ When starting a cut, always chisel away from the guideline toward the waste wood so that no splitting will occur at the edge.

☐ Never cut towards yourself with a chisel.

☐ Make the shavings thin, especially when finishing.

☐ Examine the grain of the wood to see which way it runs. Cut with the grain to sever the fibers and leaves the wood smooth. Cutting against the grain splits the wood and leaves it rough. This type of cut cannot be controlled.

Safe and proper chiseling and planing techniques will help you do a more professional job in refinishing your hardwood floor. That's it! You've learned to understand, plan, install, finish, maintain, and refinish hardwood floors. Not it's time to sit back and enjoy!

Appendices

Appendix A

Hardwood Floor Game Markings and Tables

Figures A-1 through A-9 are specific markings that can be made on finished hardwood floors in gymnasiums and sports areas, courtesy of the Maple Flooring Manufacturers Association. Tables A-1 through A-4 will help you in fastening and measuring hardwood floors.

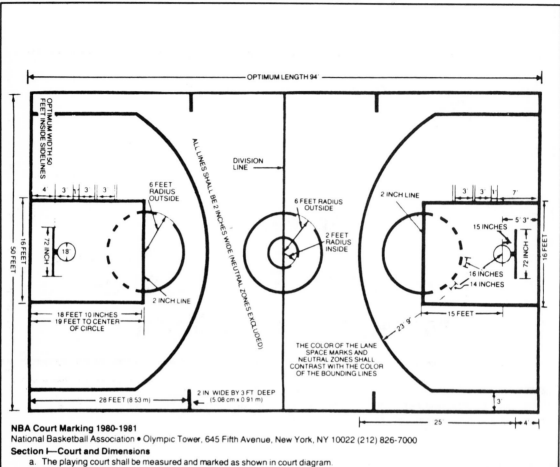

NBA Court Marking 1980-1981

National Basketball Association • Olympic Tower, 645 Fifth Avenue, New York, NY 10022 (212) 826-7000

Section I—Court and Dimensions

a. The playing court shall be measured and marked as shown in court diagram.

b. A free throw lane shall be marked at each end of the court with dimensions and markings as shown on court diagram. All boundary lines are part of the lane; lane space marks and neutral zone marks are **not**. The color of the lane space marks and neutral zones shall contrast with the color of the boundary lines. The areas identified by the lane space markings are two inches by thirty-six and the neutral zone marks are twelve inches by thirty-six inches.

c. A free throw line, 2 inches wide, shall be drawn across each of the circles indicated in court diagram. It shall be parallel to the end line and shall be 15 feet from the plane of the face of the backboard.

d. Three-point field goal area which has paralell lines three feet from the sidelines, extending from the baseline, and an arc of 23 feet-nine inches from the middle of the basket which intersects the parallel lines.

e. Four hash marks shall be drawn (two inches wide) perpendicular to the side line on each side of the court and 28 feet from the base line. The hash mark shall extend three feet onto the court.

NOTE The National Basketball Association has issued new regulations regarding defensive play and coaches boxes. These regulations affect game line markings. Consult the NBA (212) 826-7000 for specific guidance.

Fig. A-1. NBA basketball court marking.

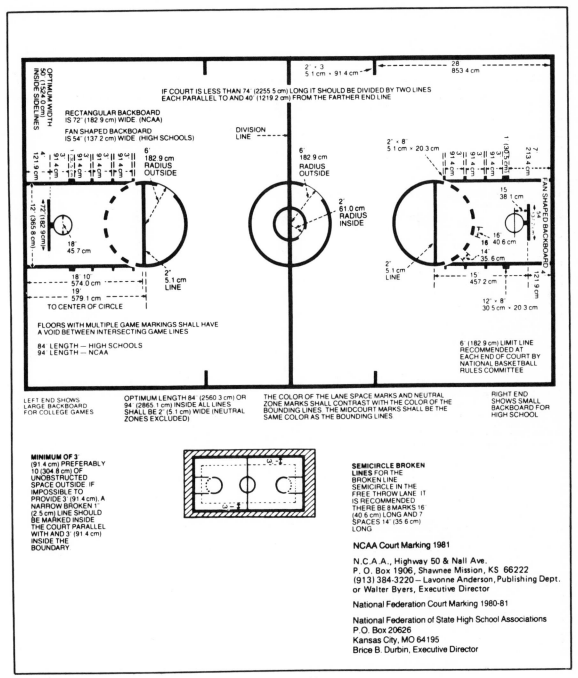

OPTIMUM WIDTH 50' (1524.0 cm) INSIDE SIDELINES

IF COURT IS LESS THAN 74' (2255.5 cm) LONG IT SHOULD BE DIVIDED BY TWO LINES EACH PARALLEL TO AND 40' (1219.2 cm) FROM THE FARTHER END LINE

2' × 3'
5.1 cm × 91.4 cm

28'
853.4 cm

RECTANGULAR BACKBOARD IS 72" (182.9 cm) WIDE. (NCAA)

FAN SHAPED BACKBOARD IS 54" (137.2 cm) WIDE (HIGH SCHOOLS)

DIVISION LINE

6'
182.9 cm
RADIUS OUTSIDE

6'
182.9 cm
RADIUS OUTSIDE

2'
61.0 cm
RADIUS INSIDE

2' × 8'
5.1 cm × 20.3 cm

1' (30.5 cm)

7'
213.4 cm

15'
38.1 cm

FAN SHAPED BACKBOARD 4'

54' (137.2 cm)

16'
40.6 cm

14'
35.6 cm

2'
5.1 cm
LINE

72" (182.9 cm)

12' (365.8 cm)

18"
45.7 cm

18' 10"
574.0 cm
19'
579.1 cm
TO CENTER OF CIRCLE

2'
5.1 cm
LINE

15'
457.2 cm

121.9 cm

12' × 8'
30.5 cm × 20.3 cm

FLOORS WITH MULTIPLE GAME MARKINGS SHALL HAVE A VOID BETWEEN INTERSECTING GAME LINES

84' LENGTH — HIGH SCHOOLS
94' LENGTH — NCAA

6' (182.9 cm) LIMIT LINE RECOMMENDED AT EACH END OF COURT BY NATIONAL BASKETBALL RULES COMMITTEE

LEFT END SHOWS LARGE BACKBOARD FOR COLLEGE GAMES

OPTIMUM LENGTH 84' (2560.3 cm) OR 94' (2865.1 cm) INSIDE ALL LINES SHALL BE 2' (5.1 cm) WIDE (NEUTRAL ZONES EXCLUDED)

THE COLOR OF THE LANE SPACE MARKS AND NEUTRAL ZONE MARKS SHALL CONTRAST WITH THE COLOR OF THE BOUNDING LINES. THE MIDCOURT MARKS SHALL BE THE SAME COLOR AS THE BOUNDING LINES

RIGHT END SHOWS SMALL BACKBOARD FOR HIGH SCHOOL

MINIMUM OF 3' (91.4 cm) PREFERABLY 10' (304.8 cm) OF UNOBSTRUCTED SPACE OUTSIDE. IF IMPOSSIBLE TO PROVIDE 3' (91.4 cm), A NARROW BROKEN 1" (2.5 cm) LINE SHOULD BE MARKED INSIDE THE COURT PARALLEL WITH AND 3' (91.4 cm) INSIDE THE BOUNDARY

SEMICIRCLE BROKEN LINES FOR THE BROKEN LINE SEMICIRCLE IN THE FREE THROW LANE IT IS RECOMMENDED THERE BE 8 MARKS 16" (40.6 cm) LONG AND 7 SPACES 14" (35.6 cm) LONG

NCAA Court Marking 1981

N.C.A.A., Highway 50 & Nall Ave.
P. O. Box 1906, Shawnee Mission, KS 66222
(913) 384-3220 — Lavonne Anderson, Publishing Dept.
or Walter Byers, Executive Director

National Federation Court Marking 1980-81

National Federation of State High School Associations
P.O. Box 20626
Kansas City, MO 64195
Brice B. Durbin, Executive Director

Fig. A-2. High school and NCAA college basketball court markings.

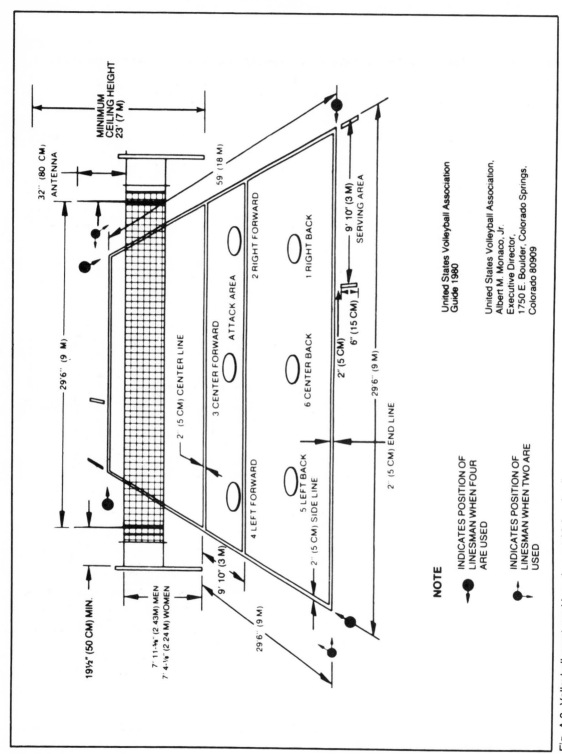

MINIMUM CEILING HEIGHT 23' (7 M)

32'' (80 CM) ANTENNA

59 (18 M)

29'6'' (9 M)

2'' (5 CM) CENTER LINE

3 CENTER FORWARD

ATTACK AREA

2 RIGHT FORWARD

1 RIGHT BACK

6 CENTER BACK

4 LEFT FORWARD

5 LEFT BACK

2'' (5 CM) SIDE LINE

9' 10' (3 M) SERVING AREA

2'' (5 CM)

6'' (15 CM)

29'6'' (9 M)

2'' (5 CM) END LINE

19½'' (50 CM) MIN.

7' 11-¾'' (2.43M) MEN

7' 4-⅞'' (2.24 M) WOMEN

9' 10' (3 M)

29'6'' (9 M)

United States Volleyball Association Guide 1980

United States Volleyball Association,
Albert M. Monaco, Jr.
Executive Director,
1750 E. Boulder, Colorado Springs,
Colorado 80909

NOTE

INDICATES POSITION OF LINESMAN WHEN FOUR ARE USED

INDICATES POSITION OF LINESMAN WHEN TWO ARE USED

Fig. A-3. Volleyball court markings (except high school).

HIGH SCHOOL VOLLEYBALL COURT RULES 1980-81 National Federation of State High School Associations

SECTION 2 The Court and Markings

*ART. 1 . . The Court shall be 60 feet long and 30 feet wide, including the outer edges of the boundary lines. An area above the court which shall be clear of any obstruction and at least 30 feet high is recommended.

ART. 2 . . It is recommended all boundary lines shall be of one clearly visible color.

ART. 3 . . Boundary lines shall be 2 inches wide and at least 6 feet from walls or obstacles. The end lines are the boundary lines on the short sides of the court. The sidelines are the boundary lines on the long sides of the court.

*ART. 4 . . A centerline, 4 inches wide, parallel to and equidistant from the end lines, shall separate the court into two playing areas.

ART. 5 . . A spiking line, 2 inches wide, shall be drawn across each playing area from sideline to sideline, the midpoint of which shall be 10 feet from the midpoint of the center line and parallel to it.

*ART. 6 . . A serving area shall be provided beyond each end line. Each shall be demarcated by two lines 6 inches long by 2 inches wide, beginning 8 inches beyond the end line and drawn perpendicular to it, one on the extension of the right sideline and the other 10 feet left of the extension. Each serving area shall be minimum of 6 feet in depth. In the event that such a space is not provided by the physical plant, the serving area shall extend into the court to whatever distance necessary to provide the minimum depth and be so marked.

*See Situations and Rulings

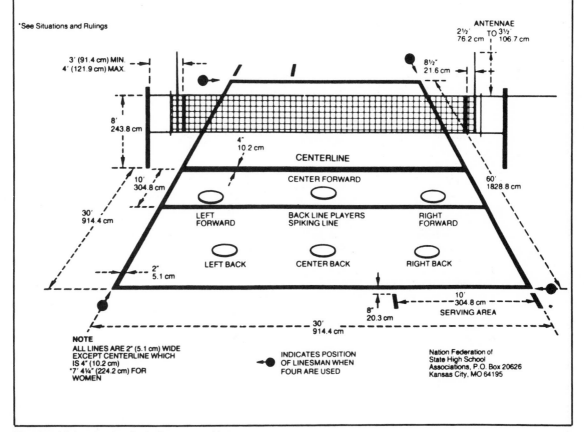

Fig. A-4. High school volleyball court markings.

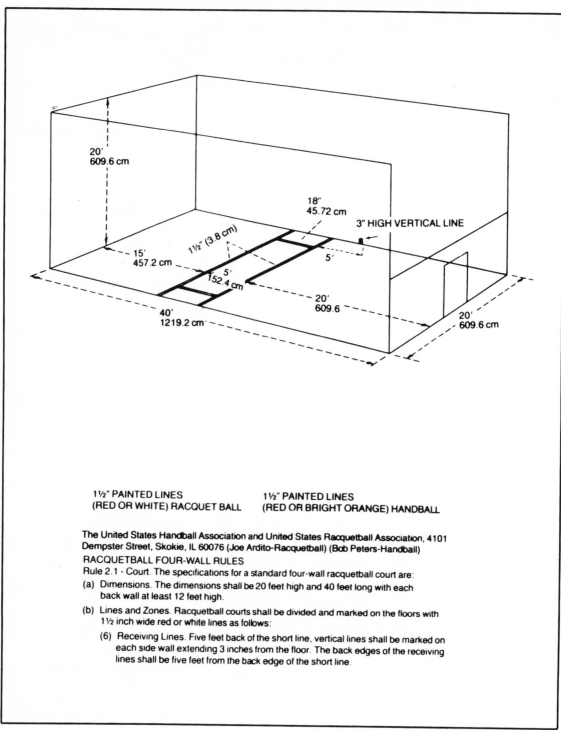

**1½" PAINTED LINES
(RED OR WHITE) RACQUET BALL**

**1½" PAINTED LINES
(RED OR BRIGHT ORANGE) HANDBALL**

The United States Handball Association and United States Racquetball Association, 4101
Dempster Street, Skokie, IL 60076 (Joe Ardito-Racquetball) (Bob Peters-Handball)

RACQUETBALL FOUR-WALL RULES

Rule 2.1 - Court. The specifications for a standard four-wall racquetball court are:

(a) Dimensions. The dimensions shall be 20 feet high and 40 feet long with each
back wall at least 12 feet high.

(b) Lines and Zones. Racquetball courts shall be divided and marked on the floors with
1½ inch wide red or white lines as follows:

    (6) Receiving Lines. Five feet back of the short line, vertical lines shall be marked on
    each side wall extending 3 inches from the floor. The back edges of the receiving
    lines shall be five feet from the back edge of the short line.

Fig. A-5. Four-wall handball and racquetball court markings.

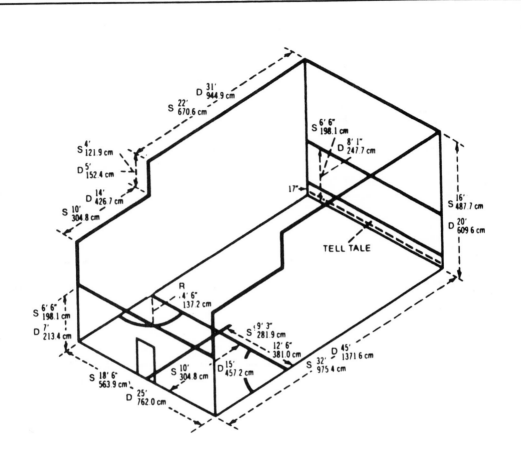

TELL TALE

Markings are inside dimensions as lines are out of bounds (except in service areas).

1980 United States Squash Racquets Association
211 Ford Road, Bala-Cynwyd, PA 19004
(215) 667-4006 Mr. Kingsley, Executive Director

Dimensions shown above preceeded by "S" are for singles court; by a "D" are for a doubles court.

Dimensions for red lines to be painted on floor and front wall are to center of 1" (2.5 cm) wide line.

Fig. A-6. Standard squash court markings.

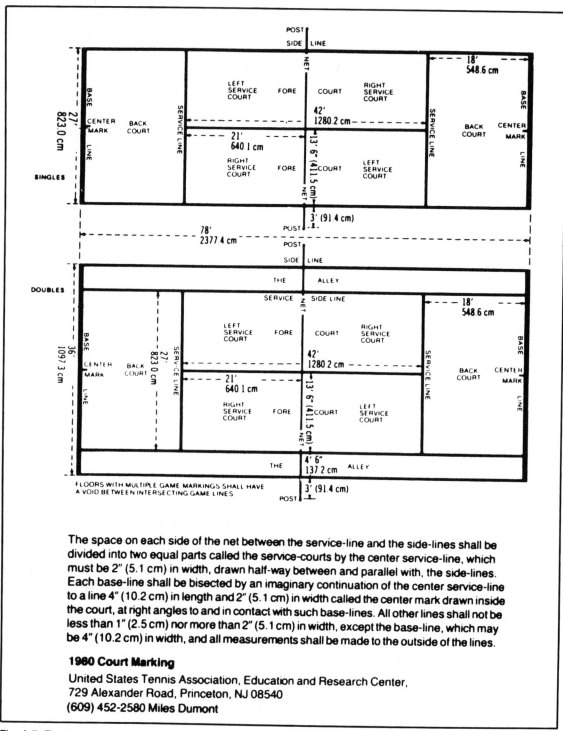

The space on each side of the net between the service-line and the side-lines shall be divided into two equal parts called the service-courts by the center service-line, which must be 2" (5.1 cm) in width, drawn half-way between and parallel with, the side-lines. Each base-line shall be bisected by an imaginary continuation of the center service-line to a line 4" (10.2 cm) in length and 2" (5.1 cm) in width called the center mark drawn inside the court, at right angles to and in contact with such base-lines. All other lines shall not be less than 1" (2.5 cm) nor more than 2" (5.1 cm) in width, except the base-line, which may be 4" (10.2 cm) in width, and all measurements shall be made to the outside of the lines.

1980 Court Marking

United States Tennis Association, Education and Research Center,
729 Alexander Road, Princeton, NJ 08540
(609) 452-2580 Miles Dumont

Fig. A-7. Tennis court markings.

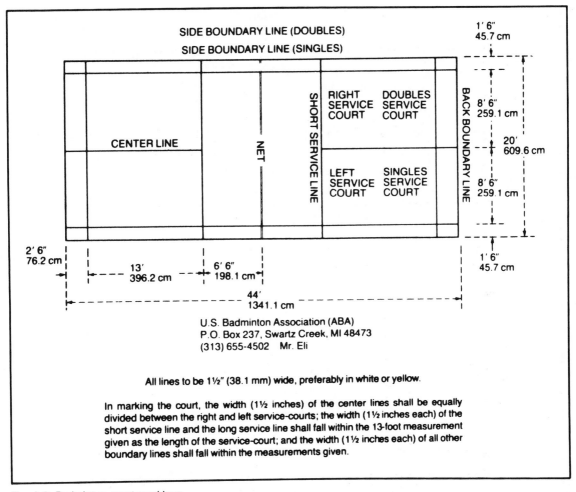

U.S. Badminton Association (ABA)
P.O. Box 237, Swartz Creek, MI 48473
(313) 655-4502   Mr. Eli

All lines to be 1½" (38.1 mm) wide, preferably in white or yellow.

In marking the court, the width (1½ inches) of the center lines shall be equally divided between the right and left service-courts; the width (1½ inches each) of the short service line and the long service line shall fall within the 13-foot measurement given as the length of the service-court; and the width (1½ inches each) of all other boundary lines shall fall within the measurements given.

Fig. A-8. Badminton court markings.

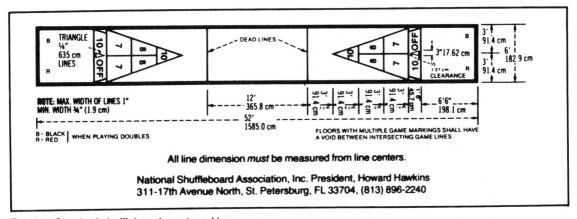

National Shuffleboard Association, Inc. President, Howard Hawkins
311-17th Avenue North, St. Petersburg, FL 33704, (813) 896-2240

Fig. A-9. Standard shuffleboard court markings.

Table A-1. Nail Sizes.

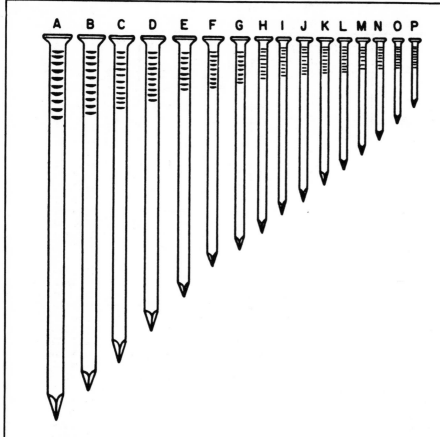

COMMON WIRE NAILS

		LENGTH AND GAUGE		APPROXIMATE NUMBER TO POUND
SIZE		INCHES	NUMBER	
A	60d	6	2	11
B	50d	5½	3	14
C	40d	5	4	18
D	30d	4½	5	24
E	20d	4	6	31
F	16d	3½	7	49
G	12d	3¼	8	63
H	10d	3	9	69
I	9d	2¾	10¼	96
J	8d	2½	10¼	106
K	7d	2¼	11½	161
L	6d	2	11½	181
M	5d	1¾	12½	271
N	4d	1½	12½	316
O	3d	1¼	14	568
P	2d	1	15	876

Table A-2. Conversion: Feet to Meters.

Feet	Meters	Feet	Meters	Feet	Meters
1	0.3048	35	10.6680	68	20.7264
2	0.6096	36	10.9728	69	21.0312
3	0.9144	37	11.2776	70	21.3360
4	1.2192	38	11.5824	71	21.6408
5	1.5240	39	11.8872	72	21.9456
6	1.8288	40	12.1920	73	22.2504
7	2.1336	41	12.4968	74	22.5552
8	2.4384	42	12.8016	75	22.8600
9	2.7432	43	13.1064	76	23.1648
10	3.0480	44	13.4112	77	23.4696
11	3.3528	45	13.7160	78	23.7744
12	3.6576	46	14.0208	79	24.0792
13	3.9624	47	14.3256	80	24.3840
14	4.2672	48	14.6304	81	24.6888
15	4.5720	49	14.9352	82	24.9936
16	4.8768	50	15.2400	83	25.2984
17	5.1816	51	15.5448	84	25.6032
18	5.4864	52	15.8496	85	25.9080
19	5.7912	53	16.1544	86	26.2128
20	6.0960	54	16.4592	87	26.5176
21	6.4008	55	16.7640	88	26.8224
22	6.7056	56	17.0688	89	27.1272
23	7.0104	57	17.3736	90	27.4320
24	7.3152	58	17.6784	91	27.7368
25	7.6200	59	17.9832	92	28.0416
26	7.9248	60	18.2880	93	28.3464
27	8.2296	61	18.5928	94	28.6512
28	8.5344	62	18.8976	95	28.9560
29	8.8392	63	19.2024	96	29.2608
30	9.1440	64	19.5072	97	29.5656
31	9.4488	65	19.8120	98	29.8704
32	9.7536	66	20.1168	99	30.1752
33	10.0584	67	20.4216	100	30.4800
34	10.3632				

Table A-3. Conversion: Millimeters to Inches.

Milli-meters	Inches	Milli-meters	Inches	Milli-meters	Inches	Milli-meters	Inches
1	0.039370	26	1.023622	51	2.007874	76	2.992126
2	0.078740	27	1.062992	52	2.047244	77	3.031496
3	0.118110	28	1.102362	53	2.086614	78	3.070866
4	0.157480	29	1.141732	54	2.125984	79	3.110236
5	0.196850	30	1.181102	55	2.165354	80	3.149606
6	0.236220	31	1.220472	56	2.204724	81	3.188976
7	0.275591	32	1.259843	57	2.244094	82	3.228346
8	0.314961	33	1.299213	58	2.283465	83	3.267717
9	0.354331	34	1.338583	59	2.322835	84	3.307087
10	0.393701	35	1.377953	60	2.362205	85	3.346457
11	0.433071	36	1.417323	61	2.401575	86	3.385827
12	0.472441	37	1.456693	62	2.440945	87	3.425197
13	0.511811	38	1.496063	63	2.480315	88	3.464567
14	0.551181	39	1.535433	64	2.519685	89	3.503937
15	0.590551	40	1.574803	65	2.559055	90	3.543307
16	0.629921	41	1.614173	66	2.598425	91	3.582677
17	0.669291	42	1.653543	67	2.637795	92	3.622047
18	0.708661	43	1.692913	68	2.677165	93	3.661417
19	0.748031	44	1.732283	69	2.716535	94	3.700787
20	0.787402	45	1.771654	70	2.755906	95	3.740157
21	0.826772	46	1.811024	71	2.795276	96	3.779528
22	0.866142	47	1.850394	72	2.834646	97	3.818898
23	0.905512	48	1.889764	73	2.874016	98	3.858268
24	0.944882	49	1.929134	74	2.913386	99	3.897638
25	0.984252	50	1.968504	75	2.952756	100	3.937008

Table A-4. Fractions and Decimal Equivalents.

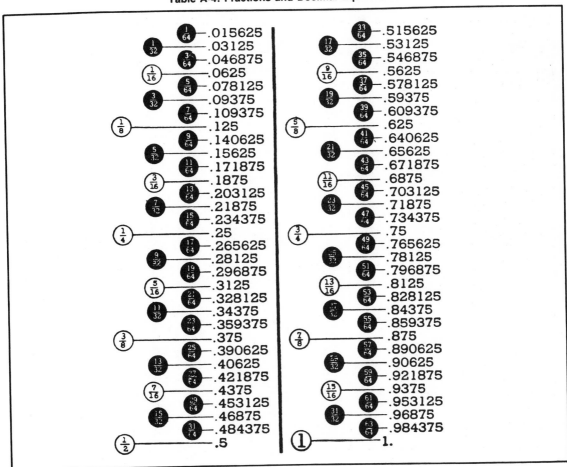

	$\frac{1}{64}$ .015625		$\frac{33}{64}$ .515625	
$\frac{1}{32}$	.03125	$\frac{17}{32}$	.53125	
	$\frac{3}{64}$ .046875		$\frac{35}{64}$ .546875	
$\frac{1}{16}$	.0625	$\frac{9}{16}$	.5625	
	$\frac{5}{64}$ .078125		$\frac{37}{64}$ .578125	
$\frac{3}{32}$	.09375	$\frac{19}{32}$	.59375	
	$\frac{7}{64}$ .109375		$\frac{39}{64}$ .609375	
$\frac{1}{8}$	.125	$\frac{5}{8}$	.625	
	$\frac{9}{64}$ .140625		$\frac{41}{64}$ .640625	
$\frac{5}{32}$	.15625	$\frac{21}{32}$	.65625	
	$\frac{11}{64}$ .171875		$\frac{43}{64}$ .671875	
$\frac{3}{16}$	.1875	$\frac{11}{16}$	.6875	
	$\frac{13}{64}$ .203125		$\frac{45}{64}$ .703125	
$\frac{7}{32}$	.21875	$\frac{23}{32}$	.71875	
	$\frac{15}{64}$ .234375		$\frac{47}{64}$ .734375	
$\frac{1}{4}$	.25	$\frac{3}{4}$	.75	
	$\frac{17}{64}$ .265625		$\frac{49}{64}$ .765625	
$\frac{9}{32}$	.28125	$\frac{25}{32}$	.78125	
	$\frac{19}{64}$ .296875		$\frac{51}{64}$ .796875	
$\frac{5}{16}$	.3125	$\frac{13}{16}$	.8125	
	$\frac{21}{64}$ .328125		$\frac{53}{64}$ .828125	
$\frac{11}{32}$	.34375	$\frac{27}{32}$	.84375	
	$\frac{23}{64}$ .359375		$\frac{55}{64}$ .859375	
$\frac{3}{8}$	.375	$\frac{7}{8}$	.875	
	$\frac{25}{64}$ .390625		$\frac{57}{64}$ .890625	
$\frac{13}{32}$	.40625	$\frac{29}{32}$	.90625	
	$\frac{27}{64}$ .421875		$\frac{59}{64}$ .921875	
$\frac{7}{16}$	.4375	$\frac{15}{16}$	.9375	
	$\frac{29}{64}$ .453125		$\frac{61}{64}$ .953125	
$\frac{15}{32}$	.46875	$\frac{31}{32}$	.96875	
	$\frac{31}{64}$ .484375		$\frac{63}{64}$ .984375	
$\frac{1}{2}$	.5	1	1.	

Appendix B

Suppliers

There are numerous manufacturers and suppliers of hardwood flooring throughout the United States and Canada. Many belong to one or more of three trade associations:

American Parquet Association (APA)
1650 Union National Plaza
Little Rock, AR 72201

National Oak Flooring
Manufacturers Association (NOFMA)
804 Sterick Building
Memphis, TN 38103

Maple Flooring Manufacturers Association
Sperry Univac Plaza
Suite 720-South
Chicago, IL 60631

Together, these trade associations sponsor the Hardwood Flooring Installation School. It is open to anyone interested in the wood flooring trades:

Manufacturers, distributors, dealers, designers, architects, builders, installers, and their employees. For further information on this school, contact any one of the trade associations.

Members of these hardwood flooring trade associations and other industry leaders include:

AGA, Inc. (MFMA)
Box 246
Amasa, MI 49903

A.J. King Lumber Co., Inc. (APA)
Box 272
Sevierville, TN 37862

Applied Radiant Energy Corp. (APA)
2432 Lakeside Drive
Lynchburg, VA 24501

Arkansas Oak Flooring Co. (NOFMA)
Box 7227
Pine Bluff, AR 71611

The Burruss Company (NOFMA)
Box 129
Lynchburg, VA 24505

Conner Forest Industries (MFMA)
Box 847
Wausau, WI 54401

Cousineau Lumber Company (NOFMA)
Box 260
Strong, ME 04983

Cumberland Lumber & Manufacturing Co.
(NOFMA)
Box 450
McMinnville, TN 37110

DeSoto Hardwood Flooring Company (NOFMA)
Box 1201
Memphis, TN 38101

Dixon Lumber Co., Inc.
Box 420
Galax, VA 24333

Harris-Tarkett, Inc. (NOFMA, MFMA)
Box 300
Johnson City, TN 37601

Horner Flooring Co. (MFMA)
Dollar bay, MI 49922

McMinnville Manufacturing Co. (NOFMA)
Box 151
McMinnville, TN 37110

Memphis Hardwood Flooring Co. (NOFMA)
Box 7253
Memphis, TN 38107

Miller & Company, Inc. (NOFMA)
Box 779
Selma, AL 36701

Missouri Hardwood Flooring Co. (NOFMA)
114 N. Gay Ave.
St. Louis, MO 63105

Monticello Flooring & Lumber Co. (NOFMA)
Box 637
Monticello, KY 42633

Oregon Lumber Company
P.O. Box 711
Lake Oswego, OR 97034

R.C. Owen Company (NOFMA)
Box 533
Hopkinsville, KY 42240

Partee Flooring Mill (NOFMA)
Box 667
Magnolia, AR 71753

Peace Flooring Company, Inc. (APA)
Box 87
Magnolia, AR 71753

Pennwood Products Co. (APA)
Box 180
East Berlin, PA 17316

P.W. Plumly Lumber Corp. (NOFMA)
Box 2280
Winchester, VA 22601

Potlatch Corporation (NOFMA)
Box 390
Warren, AR 71671

Robbins, Inc. (MFMA)
3626 Roundbottom Road
Cincinnati, OH 45244

Searcy Flooring Inc.,
Division of Robbins, Inc., (NOFMA)
Box 906
Searcy, AR 92143

Smith Flooring Inc. (NOFMA)
Box 99
Mt. View, MO 65548

Sykes Flooring Products, Masonite Corp., (APA)
Box 999
Warren, AR 71671

Thomson Oak Flooring Co. (NOFMA)
Rt. #1, Box 6
Thomson, GA 30824

Zickgraf Hardwood Company (NOFMA)
Box 230
Franklin, NC 28734

Glossary

air-dried lumber—Lumber that has been piled in yards or sheds for any length of time. For the United States as a whole, the minimum moisture content of thoroughly air-dried lumber is 12 to 15 percent; the average is somewhat higher. In the South, the source of most American hardwoods, air-dried lumber may be no lower than 19 percent.

anchor bolts—Bolts to secure a wooden sill plate to concrete or masonry wall or floor.

asphalt—Most native asphalt is a residue from evaporated petroleum. It is insoluble in water but soluble in gasoline and melts when heated. Used widely in building for waterproofing floors, flooring tile, roof covers, exterior walls, etc.

backband—A simple molding sometimes used around the outer edge of plain rectangular casing as a decorative feature.

backfill—The replacement of excavated earth into a trench around and against a basement foundation.

baseboard—A board placed against the wall around a room next to the floor to finish properly between floor and wall. Also called a base.

base molding—Molding used to trim the upper edge of interior baseboard.

base shoe—Molding used next to the floor on interior baseboards. Sometimes called a carpet strip.

beam—A structural member transversely supporting a load.

bearing partition—A partition that supports any vertical load in addition to its own weight.

bearing wall—A wall that supports any vertical load in addition to its own weight.

blind nailing—Nailing in such a way that the nailheads are not visible on the face of the work; usually at the tongue of matched boards.

block flooring—Wood flooring made of blocks (square or rectangular pieces of wood), rather than strips.

bodied linseed oil—Linseed oil that has been thickened in viscosity by suitable processing with heat or chemicals.

boiled linseed oil—Linseed oil in which enough lead, manganese, or cobalt salts have been incorporated to make the oil harden more rapidly when spread in thin coatings.

brace—An inclined piece of framing lumber applied to a wall or floor to stiffen the structure.

bridging—Small wood or metal members that are inserted in a diagonal position between the floor joists at midspan to act both as tension and compression members for the purpose of bracing the joists and spreading the action of loads.

butt joint—The junction of the ends of two timbers or flooring boards in a square-cut joint.

condensation—Beads or drops of water (and frequently frost in extremely cold weather) that accumulate on the inside of the exterior covering of a building when warm, moisture-laden air from the interior reaches a point where the temperature no longer permits the air to sustain the moisture it holds.

cut-in brace—Nominal 2-inch-thick members, usually 2-×-4s, cut in between each stud diagonally.

cove molding—A molding with a concave face used as trim or to finish interior corners.

crawl space—A shallow space between the living quarters of a basementless house, normally enclosed by the foundation wall.

cross bridging—Diagonal bracing between adjacent floor joists placed near the center of the joist span to prevent joists from twisting.

crown molding—A molding used on cornice or wherever an interior angle is to be covered.

d—Penny; as applied to nails, it originally indicated the price per hundred. The term now serves as a measure of nail length.

dado—A rectangular groove across the width of a board or plank.

direct nailing—Nailing perpendicular to the initial surface or to the junction of the pieces joined. Also called face nailing.

drywall—Interior covering material, such as gypsum board or plywood, which is applied in large sheets or panels.

drywall construction—A type of construction in which the interior wall finish is applied in a dry condition, generally in the form of sheet materials or wood paneling, as contrasted to wet plaster.

fascia—A flat board or face used sometimes by itself, but usually in combination with moldings, often located at the outer face of the cornice.

filler (wood)—A heavily pigmented preparation used for filling and leveling off the pores in open-pored woods.

footing—A masonry section, usually concrete, in a rectangular form wider than the bottom of the foundation wall or pier it supports.

foundation—The supporting portion of a structure below the first floor construction, or below grade, including the footings.

frost line—The depth of frost penetration in soil. This depth varies in different parts of the country.

girder—A large principal beam of wood or steel used to support concentrated loads at isolated points along its length.

grain—The direction, size, arrangement, appearance, or quality of the fibers in wood.

grain, edge—Lumber that has been sawed parallel to the pith of the log and approximately at right angles to the growth rings; i.e., the rings form an angle of less than 45 degrees with the surface of the piece. Also called vertical grain.

grain, flat—Lumber that has been sawed parallel to the pith of the log and approximately tangent to the growth rings; i.e., the rings form an angle of less than 45 degrees with the surface of the piece.

hardwood—Wood from deciduous trees, such as maple, oak, and cherry, known for its hard and compact substance.

header—A beam placed perpendicular to joists and to which joists are nailed in framing.

heartwood—The wood extending from the pith to the sapwood, the cells of which no longer participate in the life process of the tree.

insulation—Any material high in resistance to heat transmission that, when placed in the floors, walls, or ceiling of a structure, will reduce the rate of heat flow.

joint—The space between the adjacent surfaces of two members or components joined and held together by nails, glue, cement, mortar, or other means.

joist—One of a series of parallel beams, usually 2 inches in thickness, used to support floor and ceiling loads and supported in turn by larger beams, girders, or bearing walls.

kiln-dried lumber—Lumber that has been kiln-dried, often to a moisture content of 6 to 12 percent. Common varieties of softwood lumber, such as framing lumber, are dried to a somewhat higher moisture content.

knot—In lumber, the portion of a branch or limb of a tree that appears on the edge or face of a piece.

lumber—The wood product of the sawmill and the planing mill not further manufactured other than by sawing, resawing, and passing lengthwise through a standard planing machine, crosscutting to length and matching. Boards are lumber less than 2 inches thick and 2 or more inches wide. Dimension lumber is hard lumber from 2 to, but not including, 5 inches thick and 2 or more inches wide.

matched lumber—Lumber that is dressed and shaped on one edge in a grooved pattern and on the other edge in a tongued pattern.

millwork—Generally, all building materials made of finished wood and manufactured in millwork plants and planing mills, but not flooring, ceiling, or siding.

miter joint—The joint of two pieces of an angle that bisects the joining angle. For example, the miter joint at the corner of a floor (90 degrees) is made at a 45-degree angle.

molding—A wood strip having a curved or projecting surface used for decorative purposes.

mortise—A slot cut into a board, plank, or timber, usually edgewise, to receive the tenon of another board, plank, or timber to form a joint.

nonbearing wall—A wall supporting no load other than its own weight.

notch—A crosswise rabbet at the end of a board.

O.C.—On center: the measurement of spacing for studs, rafters, joists, and the like in a building from the center of one member to the center of the next.

paper, building—A general term for papers, felts, and similar sheet materials used in buildings without reference to their properties or uses.

parquet—Inlaid mosaic of wood, used especially for floors. Technically, a floor of *parquetry* or wood mosaics in individual or block pieces is called a parquet floor.

pier—A column of masonry, usually rectangular in horizontal cross section, used to support other structural members, such as flooring joists.

pitch pocket—An opening extending parallel to the annual rings of growth that usually contains or has contained either solid or liquid pitch.

pith—The small, soft core at the original center of a tree around which wood formation takes place.

plywood—A piece of wood made of three or more layers of veneer joined with glue and usually laid with the grain of adjoining plies at right angles. Almost always an odd number of plies are used to provide balanced construction.

pores—Wood cells of comparatively large diameter that have open ends and are set one above the other to form continuous tubes. The openings of the vessels on the surface of a piece of wood are referred to as pores.

quarter round—A small molding that has a cross section in the shape of a quarter circle.

rabbet—A rectangular longitudinal groove cut in the corner edge of a board or plank.

sapwood—The outer zone of wood next to the bark. In the living tree it contains some living cells, as well as dead and dying cells. In most species it is lighter colored than the heartwood. In all species it is lacking in decay resistance.

screed—A small strip of wood laid under the flooring or subflooring and above a concrete slab to separate the two.

sealer—A finishing material, either clear or pigmented, that is usually applied directly over uncoated wood to seal the surface.

shellac—A transparent coating made by dissolving lac, a resinous secretion of the lac bug, in alcohol.

sleeper—Usually a wood member embedded in concrete, as in a floor, that serves to support and to fasten subflooring or flooring.

softwood—Wood from conifer trees such as Douglas fir, hemlock, and southern pine.

soil cover—A light covering of plastic film, roll roofing, or similar material used over the soil in crawl spaces of buildings to minimize moisture permeation of the area.

sole plate—The bottom horizontal member of a frame wall.

solid bringing—A solid member placed between adjacent floor joists near the center of the span to prevent the joists from twisting.

span—The distance between structural supports such as walls, columns, piers, beams, girders, and trusses.

story—The part of a building between any floor and the floor or roof above it.

stringer—A timber or other support for cross members in floors or ceilings.

strip flooring—Wood flooring consisting of narrow, matched strips, often of tongue-and-groove hardwood.

stud—One of a series of slender wood or metal vertical structural members placed as supporting elements in walls and partitions.

subfloor—Boards or plywood laid on joists over which a finish floor is to be laid.

termites—Insects that superficially resemble ants in size, general appearance, and habit of living in colonies; hence, they are frequently called "white ants." Subterranean termites establish themselves in buildings, not by being carried in with lumber, but by entering from ground nests after the building has been constructed. If unmolested, they eat out the woodwork, leaving a shell of sound wood to conceal their activities. Damage may proceed so far as to cause collapse of parts of a structure before discovery. There are about 56 species of termites known in the United States; but the two major ones, classified by the manner in which they attack wood, are ground-inhabitating or subterranean termites (the most common) and dry-wood termites, which are found almost exclusively along the extreme southern border and the Gulf of Mexico in the United States.

termite shield—A shield, usually of noncorrodible metal, placed in or on a foundation wall or other mass of masonry or around pipes to prevent passage of termites.

toenailing—To drive a nail at a slant with the initial surface in order to permit it to penetrate into a second member.

tongue and groove—Boards or planks machined in such a manner that there is a groove on one edge and a corresponding tongue on the other. The most popular dressing joint for hardwood flooring. Also known as *dressed and matched*.

trim—The finish materials in a building, such as baseboards or cornice moldings, applied around openings or at the floor and ceiling of rooms.

turpentine—A volatile oil used as a thinner in paints and as a solvent for varnishes. Chemically, a mixture of terpenes.

underlayment—A material placed under finish coverings, such as flooring or shingles, to provide a smooth, even surface for applying the finish.

vapor barrier—Material used to retard the movement of water vapor into floors and walls to prevent condensation in them; usually considered as having a perm value of less than 1.0. Applied separately over the warm side of exposed walls or as a part of batt or blanket insulation.

varnish—A thickened preparation of drying oil or drying oil and resin suitable for spreading on surfaces to form continuous, transparent coatings or for mixing with pigments to make enamels.

vehicle—The liquid portion of a finishing material; it consists of the binder and volatile thinners.

veneer—Thin sheets of wood made by rotary cutting or slicing of a log. Hardwood veneers are often used for flooring.

volatile thinner—A liquid that evaporates readily and is used to thin or reduce the consistency of finishes without altering the relative volumes of pigments and nonvolatile vehicles.

wane—Bark, or lack of wood from any cause, on the edge or corner of a piece of wood.

wood rays—Strips of cells extending radially within a tree and varying in height from a few cells in some species to 4 inches or more in oak. The rays serve primarily to store food and to transport it horizontally in the tree.

Index

A

adhesive, 78

B

backsaw, 26
bark, 3
base moldings, 92-96
bell face, 31
bleaching, 90
block flooring, 65-67
block flooring, installing over a slab, 57-58
block flooring, sanding, 84
block floors, repairing, 110
block plane, 117
board feet, estimating, 19
bottom plate, 39
bridging, 37
buying hardwood flooring, 19-20

C

cambium layer, 3
care of hardwood floors, general, 97
care, daily, 98-99
care, easy floor, 100-101
caring for handsaws, 27
carpenter's level, 30-31
chiseling, refinish, 118-120
circular saw, portable, 27
claw hammer, 31-32
claw, 31
cleat stairway, 39
close-grained, 3
coarse-grained, 3
colored penetrating sealers, 85

coloring the floor, 85, 87
  colored penetrating sealers, 85
  pigmented wiping stains, 85, 87
  varnish stains, 87
combination square, using, 24
common lumber, 19
compass saw, 26
concrete slab, installing oak flooring over, 67
construction, house, 15-16
construction, platform, 15
contour cutting, 94
coping joint, 94, 96
coping saw, 26
cracks, repairing floor, 107
cracks, solving, 103-104
creaking floors, repairing, 104-105
cross-grained, 3
crosscutting, 25
cutting lumber, 4-5
cutting, contour, 94

D

d, 34
daily care of hardwood floors, 98-99
dovetail saw, 26

E

edge-grain lumber, 4
embedded water, 5
estimating board feet, 19
eye, 31

F

face nailing, 7
face, bell, 31

face, plain, 31
fiber saturation point, 5
fillers, use of, 91
fine-grained, 3
finish floor, laying out the, 59-62
finish, preparing for, 84-85
finish, protecting the, 92
finish, repairing the, 103
finish, understanding your, 99
finishes, new, 114-115
finishes, types of, 87-90
  bleaches, 90
  lacquer, 89-90
  penetrating seal, 87-88
  polyurethane, 88-89
  shellac, 89
  surface finishes, 88
  varnish, 89
finishing hardwood floors, 79-96
  base moldings, 92-96
  coloring the floor, 85-87
  floor preparation, 79
  gym and roller rink floors, 91-92
  preparing for the finish, 84-85
  protecting the finish, 92
  sanding, 79-84
  stenciling, 90-91
  types of finishes, 87-90
  use of fillers, 91
flat-grain sawing, 4
floating floor installation, 69-75
floating floor system, 69
floor care, easy, 100-101
floor cracks, repairing, 107
floor joists, 37

floor plan, 17
floor preparation, 79
floor repairs, 104-110
   repairing block and parquet floors,
     110
   repairing creaking floors, 104-105
   repairing floor cracks, 107
   repairing sagging floors, 105-107
   replacing flooring boards, 108-109
floor, coloring the, 85, 87
flooring boards, replacing, 108-109
flooring patterns, 12-14
flooring, 46, 48, 49
flooring, gym, 67-68
flooring, handling and storage of,
   52-55
flooring, installing parquet, 76-78
flooring, installing plank, 75-76
flooring, plank, 62-64
flooring, sanding parquet, block, and
   similar, 84
flooring, sanding strip and plank, 84
flooring, strip, 6
floors, gym and roller rink, 91-92
floors, repairing creaking, 104-105
floors, repairing sagging, 105-107
footing, 36
foundation, 36-38, 40-46
frame, 116
framing square, using, 23-24
framing, wall, 38-39, 46-48
free water, 5

G

general care of hardwood floors, 97
girders, 37
grain, 3
gym floors, 67-68, 91-92

H

hammer, claw, 31-32
handle, 116
handling flooring, 52-55
handsaws, 25-27
   caring for, 27
   special, 26
hardwood flooring tools, 20-32
hardwood flooring, buying, 19-20
hardwood floors, daily care of, 98-99
hardwood floors, finishing, 79-96
hardwood floors, general care of, 97
hardwood floors, installing, 51-78
hardwood floors, maintaining,
   97-110
hardwood floors, planning, 17-49
hardwood floors, refinishing,
   111-120
hardwood floors, understanding,
   1-16
heartwood, 3
heel, 25, 116

house construction, 15-16

I

installation methods, 51-52
installation over concrete slabs,
   55-58
   calcium chloride test, 55
   installing block flooring, 57-58
   plywood-on-slab method, 56
   polyethylene film test, 55
   rubber mat test, 55
   Screeds method, 56
   vapor barrier, 55-56
installation over wood joists, 58-59
installation problems, 67
   oak flooring over a radiant-heated
     concrete slab, 67
   strip flooring in a wood plenum
     system, 67
installation, easier, 69
installation, floating floor, 69-75
installing block flooring over a slab,
   57-58
   polyethylene method, 57
   two-membrane asphalt felt or
     building paper method, 57-58
installing hardwood floors, 51-78
   adhesive, 78
   floating floor installation, 69-75
   gym flooring, 67-68
   handling and storage, 52-55
   installation methods, 51-52
   installing parquet flooring, 76-78
   installing plank flooring, 75-76
   laying out the finish floor, 59-62
   leaving your mark, 78
   making installation easier, 69
   over concrete slabs, 55-58
   over old flooring, 64-65
   over wood joists, 58-59
   parquet and block flooring, 65-67
   plank flooring, 62-64
   special installation problems, 67
installing parquet flooring, 76-78
installing plank flooring, 75-76

J

jack plane, 117
joint, coping, 94, 96
jointer plane, 117
jointer, 118
joints, wood, 7-12
joists, floor, 37

K

kerf, 25
keyhole saw, 26
kiln drying, 5
kiln, 5
knob, 116

L

lacquer, 89-90
lag screw, 35
laying out the finish floor, 59-62
laying over old flooring, 64-65
level, carpenter's, 30-31
lumber, common, 19
lumber, cutting and seasoning, 4-5
lumber, edge-grain, 4
lumber, planning, 17-19
lumber, select, 19
lumber, yard, 18

M

maintaining hardwood floors, 97-110
   daily care, 98-99
   easy floor care, 100-101
   floor repairs, 104-110
   general care, 97
   maintaining penetrating seals,
     99-100
   maintaining polymer finishes, 100
   maintaining special surfaces, 101
   maintaining surface finishes, 100
   removing stains, 101-103
   repairing the finish, 103
   solving cracks and squeaks,
     103-104
   understanding your finish, 99
making installation easier, 69
materials, wood flooring, 5-7
medullary rays, 3
methods, installation, 51-52
miter box, 27
molding, base, 92-96
mouth, 116

N

nailing, 32-33
nails, 33-34
new finishes, 114-115

O

oak flooring over a radiant-heated
   concrete slab, 67
oil finish, 112
old floor, sanding an, 113-114
open-grained, 3

P

parquet flooring, 65-67
parquet flooring, installing, 76-78
parquet flooring, sanding, 84
parquet floors, repairing, 110
patterns, flooring, 12-14
penetrating seal finish, 87-88
penetrating sealers, colored, 85
penetrating seals, maintaining,
   99-100
penny, 34
Phillips head, 35